XIAO

The Story
OF
RICE

outskirts
press

The Story of Rice
All Rights Reserved.
Copyright © 2022 Xiaorong Zhang
v3.0

The opinions expressed in this manuscript are solely the opinions of the author and do not represent the opinions or thoughts of the publisher. The author has represented and warranted full ownership and/or legal right to publish all the materials in this book.

This book may not be reproduced, transmitted, or stored in whole or in part by any means, including graphic, electronic, or mechanical without the express written consent of the publisher except in the case of brief quotations embodied in critical articles and reviews.

Outskirts Press, Inc.
http://www.outskirtspress.com

ISBN: 978-1-9772-5370-5

Cover Photo © 2022 www.gettyimages.com. All rights reserved - used with permission.

Outskirts Press and the "OP" logo are trademarks belonging to Outskirts Press, Inc.

PRINTED IN THE UNITED STATES OF AMERICA

DEDICATION

*To all the rice farmers and the rice researchers.
Thank you.*

Table of Contents

Prologue i
The Beginning of the Story.. 1
Becoming Rice 10
A Tale of Two Grains.. 16
Short and Sturdy 22
Hybrid Rice.. 31
Golden Rice 43
Zongzi.. 52
Guilin Rice Noodles 61
Sushi 66
A Bowl of Gumbo 75
When Rice Met Soybeans....................................... 81
Shaoxing Rice Wine 92
The Science of Making Rice.................................. 102
A Leaf of Grass 109
The Small World of a Rice Kernel 116
The Epic Journey of Carbons 121
Annual or Perennial.. 133
The Death of Grass 137

The "Root" of the Matter 145
The Book of Rice 150
Nature versus Nurture 157
Extras.. 166
From One to Billions 172
Out of Africa 179
Bella Ciao.. 190
The Stairways to Heaven....................................... 198
Kill Two-birds with One Stone 205
Naming Matters... 211
North and South 218
Reap What You Sow.. 229
Go to Space... 243
Epilogue 250
Acknowledgements 253
Selected Bibliography 254

Prologue

IN AN ENGLISH class for international students years ago, I was asked a question: If you could select one thing that best represents your country and people, what would it be? I pondered this question for a long time. I thought of the Four Great Inventions made by the ancient Chinese—gunpowder, papermaking, printing, and the compass, which are known to every Chinese person, even pre-school children. I also considered the iconic Great Wall, the extravagant Forbidden City, the exquisite blue-and-white porcelain, or the billowing silk. I even conjured up images of cuddly giant pandas and elegant bamboo forests. But in the end, I chose rice—a small, plain, and common grain.

The Chinese domesticated rice and invented rice paddies thousands of years ago, a deed that in my opinion is much more significant and far-reaching than any of those Four Great Inventions. It was an innovation that had resulted from the collective wisdom and effort of Chinese farmers. From its humble beginning as a wild grass born out of the swamp, rice has spread all over the world and now feeds more than half the globe. It has forever changed the face of the earth and shaped the history of humanity.

Chinese civilization is literally a product of rice. In addition to being an important staple food and a major source of calories, rice has considerable cultural and historical significance in China. Its food and alcohol are mostly built around rice. Its language has been greatly enriched by this grain crop. In Chinese the character for rice is 米, which had evolved from its ancient form, a pictograph of six grains attached to a stalk. The Chinese character for an agricultural field or a plot is 田, an image of rice paddies. In addition to being stand-alone characters, 米 and 田 are also used as radicals to assemble other Chinese characters. For example, 糕 is a type of cake made from rice; 粮 means staple food; and 亩 is a unit to measure the area of a land, particularly a farmland.

Moreover, thousands of Chinese phrases, idioms, metaphors, and proverbs are directly related to rice. Rice is the "bread and butter" of Chinese. "Iron rice bowl" is a metaphor for job security, as a bowl made of metal is unbreakable and guarantees to put food on the dinner table. A proud man "never bends over backwards for five bushels of rice." Chinese people never cry over spilled milk, but they "feel regret about the cooked rice." When employees are rewarded equally regardless of their contribution, they are eating from a "big pot of rice." In America you can't make an omelet without breaking a few eggs, but in China "even the best housewife cannot cook a meal without rice." To a westerner bread and water are the bare essentials for sustenance, but to a Chinese person, it is "plain rice and cheap tea." In the West you could deprive someone's livelihood by taking the bread out of his or her mouth, but in China you could destroy a person's life by "smashing his or her rice bowl." In other parts of the world, a fertile land is depicted as a land flowing with milk and honey, but in China it is a "land of rice and fish."

There is no doubt that some ethical values and daily habits have been shaped by the hard work and the seasonal rhythm of rice farming. The virtues of thrift and frugality have been drilled into Chinese children from an early age through a poem written more than a thousand years ago, and it has been the most recited and memorized poem by generations of Chinese people. It goes like this:

Hoeing weeds under the noon sun,
Beads of sweat dripping down to the soil.
Do you know that the meal in your plate,
Every grain is wrapped with toil?

Although the elegance and poetic rhythm of this verse are somewhat lost in translation, its message is still loud and clear—every single grain of rice comes from the sweat of hard labor and shouldn't be wasted. As a child I was afraid to leave a single grain in my rice bowl, because I was told it would turn into a pockmark on my face. I was also taught to be as modest and respectful as a ripe rice plant, with its head weighed down by its copious heavy grains. In Chinese language, there are no better metaphors for modesty and humility than a bending rice plant—an analogous image of a person who is productive and knowledgeable yet humble and polite.

Chinese people call themselves the "rice people," just as the descendants of the Mayans consider themselves the "corn people" and those of the Incans refer themselves as the "potato people." Being one of the "rice people" and a botanist, I am fascinated by rice. I have traveled to archaeological sites to see seven-thousand-year-old rice kernels. I have looked at the rice plant's flowers, blades, and grains under a microscope, marveling at the intricate details and micro-patterns in them. I have analyzed the secret code in the "book of rice," its genetic blueprint, trying to understand how its tens of thousands of genes are manifested physically into pearl-like grains. I have collected rice grains of many varieties in little glass jars—a myriad of shapes, colors, and opacities—staring at them in awe. I have always grown a few pots of rice plants to decorate the yard or doorway in and around my house. Best of all, I have tried many rice recipes over the years and eaten many delicious rice-based meals.

I have learned a little about rice over the years and felt an urge to share what I know. In this book I want to tell you a few stories—the history and culture of rice. It is my hope that after reading it, you will see this unassuming grain from a different perspective and understand how it has shaped our history and changed our daily lives.

The Beginning of the Story

THE YAO RIVER is a small river by any standard, running its sixty-five-mile course through the eastern part of Zhejiang province of China. Although a mere rivulet compared to the mighty Yangtze running in the same direction, Yao was a crucial link between the inland and the Maritime Silk Road in ancient times and has been an important shipping route of China for centuries. Its watershed has nurtured the bodies and minds of people who have lived there, producing some of the most influential philosophers and thinkers in Chinese history.

Along the banks of the Yao River, small villages and towns had grown like vines. In ancient times when there were no bridges, townspeople crossed the Yao River by ferryboats, carrying goods and animals with them. Although those ancient ferries have been replaced by elegant modern bridges today, the legacy

THE STORY OF RICE

has endured through the names of some villages and towns, which all end with "du," meaning a ferry dock in Chinese language.

One of those towns is called Hemu<u>du</u>, which had been a small rice-farming town for many centuries, with its land always covered with green paddies in spring and amber grains in fall. It had been a quiet town and unknown to the outside world until 1973, when the villagers stirred up that land, trying to improve the drainage of their rice fields. As they dug deeper, bits of pottery spilled out and animal bones popped up. Archaeologists were called to the scene and an excavation started right away.

Digging through more than four meters, archaeologists were able to identify four distinct cultural layers, each of which recorded a history of a bygone era. Over a period of several years, more than seven thousand artifacts and relics were excavated, from farming tools and hunting weapons to clay pottery and ivory carvings. The crown jewel of the excavation, though, was an assortment of rice remains, including husks, stalks, and kernels, well preserved in layers of sediments and debris. A life of a distant past was gradually reconstructed by archaeologists from those bits and pieces buried below the town's rice paddies. The site was dated back to 5500 BC to 3300 BC, and it was one of the earliest rice farming villages in the Yangtze River region of China.

THE BEGINNING OF THE STORY

While visiting China in the summer of 2015, our family of four arrived at a ferry dock on the south bank of the Yao River and waited to be carried to the north, where the Hemudu Archaeological Site was located. While waiting I looked around, trying to take in the environs of Hemudu. Simmering in the morning sun, the Yao River flowed calmly through foothills. Its surface was dotted with large patches of tiny green duckweeds. Small hills were emerald green with their summer vegetation. A red-roofed stilt house was standing on the opposite shore, with its lonely reflection in the river. The site and the museum built next to the site were hidden behind a dense thicket of trees and shrubs.

We were the only passengers in the ferry boat, a small traditional rowboat. The ferryman was a quiet middle-aged man standing in the prow with a long wooden oar. He was clearly oblivious to the scenery and his clamorous passengers and rowed the boat silently and mechanically. We stopped talking and listened to the rhythmic sound of the oar pushing through the water. Halfway across the river an old barge chugged along westward with a loud engine noise. For a moment it felt a little bit surreal, as if I were being ferried across a mystical river to a far shore where I was about to take a time-travel adventure.

The museum is one of a kind. It was designed and built based on the ancient Hemudu's architectural style, reconstructed from excavated pillars, beams, boards, and other architectural elements, to harmonize it with the rest of the site. It included a high foundation, a typical feature of an ancient Hemudu-style house designed to prevent flooding. The building rests on 456 pillars on which lie groups of cross beams, symbolizing the tenon and mortise technology used by the Hemudu people seven thousand years ago. The foyer is in the shape of the spreading wings of a roc, a legendary bird worshiped by the Hemudu people.

The inside of the museum was very quiet, with just a handful of visitors; all the clang of the outside world was muffled. While I was ambling in the exhibition hall that held all the relics related to the rice farming of the time, my attention was caught by the black rice kernels laid in a modern-day petri dish. I stood a long time looking at those kernels, the organic objects made by the sunlight, air, and water of thousands of years ago. These grains actually appeared golden-yellow when they were freshly unearthed and turned black as soon as exposed to the air and light. As seen from the picture taken on the spot during the excavation, those golden kernels stood out vividly in contrast with the black loam in which they were buried. Time had not changed them a bit, as they looked just like modern-day rice grains with all the morphological details, including two parallel ridges along the husk and a sharp awn on the tip. I learned later that scientists using an advanced archaeological technique positively identified four of the rice kernels recovered from

THE BEGINNING OF THE STORY

the site as wild rice grains, but the rest were domesticated rice, all dated to about seven thousand years ago. The mix of wild and domesticated rice implies that ancient Hemudu people might have started as hunter-gatherers and grew over time into farmers during their two thousand years of settlement.

Along with rice remains, many farming tools were also on display. Among them were the spade-like ploughs made of animal bones. The Hemudu people brilliantly exploited the natural structure of the shoulder blades of several large mammals and fashioned them into all kinds of shapes—straight or slanted, flat or columnar, smooth- or rough-edged—for various purposes. When fastened to a long wooden handle, the bones could be used for digging, plowing, and weeding in the rice field. Some of the tools were prick marked on both sides of the blade, almost like a modern-day sickle, for slashing stalks and harvesting grains. Tools for processing rough grains were also excavated, such as the wooden pole with an enlarged end that could be used to pound and crush kernels to remove the husk and make the rice edible.

In a separate exhibition hall rows and rows of pottery were displayed on tiered shelves. The dim light cast the distorted shadows of the pottery on the wall behind them, giving abstract images. They were the pots, pitchers, streamers, plates, jars, cups, and bowls of the Stone Age. Those pottery wares were made with clay tempered with charcoal powder and organic matter, such as the leaf, stem, and hull of rice or other plants. They were carbonized once fired, giving a black, earthy, and porous appearance. Several on display were cord-marked or garnished with paintings or carvings.

One of the pots drew my attention. It had a T-shaped plant carved onto it, with a thick stalk and a bowing head laden with grains, a simple and vivid image of a ripe rice plant. I read from one of the figure legends that the burned rice crusts had been found in some of the pots and pottery fragments excavated at the site—a piece of tangible evidence that rice had been boiled or simmered in those pots. My thoughts wandered. Cooking rice must have been a great culinary challenge to our ancestors. Unlike wheat flour, rice flour does not contain gluten, so it resists turning into malleable dough. If the Hemudu people tried to grind rice grains into flour, the bread made of pure rice flour would have been brittle and broken apart easily when baked over a fire or under the sun. Cooks would have had to resort to cooking the whole grains in water, which meant they had a need for heat- and leak-proof cooking pots as well as bowls for serving rice. There is no doubt that rice must have played a role in motivating and compelling the Hemudu people to invent pottery that would help them cook rice, put rice onto a table, and put it into their mouths. Domestication of rice and invention of pottery had to coincide and go hand-in-hand.

THE BEGINNING OF THE STORY

**The image of a rice plant carved onto a piece of pottery excavated from the Hemudu Site.
(Courtesy of the Hemudu Site Museum)**

Without any written record, we have to imagine and piece together the life of a distant past from the objects excavated, deciphering the messages that these objects convey and interpreting how people lived in that era. The ancient terrain of Hemudu was reconstructed by scientists; it was a much warmer and wetter place thousands of years ago with a tropical climate—an ideal habitat for growing rice. The ecological landscape was a mix of forests, swamps, meadows, plains, and hills, veined by pristine rivers and streams. It was a rich world populated by tropical flora and fauna with a multitude of mammals, birds, fishes, and trees and grasses, as revealed by unearthed fossils and remains. Standing in front of the artistically rendered landscape, with a leap of poetic imagination I saw a utopian society in which people

THE STORY OF RICE

lived harmoniously and idyllically with nature and with each other. The Hemudu people were rice farmers as well as hunters and gatherers. They foraged for all kinds of wild plants, including acorns, melons, peaches, kiwifruits, and berries. They also fished and hunted, as evidenced by the fishing and hunting tools they left behind, such as bone harpoons, bows, and arrowheads. In addition to obtaining their sustenance, they also created all kinds of art—ivory carvings, jade ornaments, clay figurines, and lacquered bowls. They decorated their pottery with geometric shapes, symbols, and patterns. They even created flutes and whistles from the femurs of animals and drums from the wood, adding new melody and rhythm to their lives.

The existence and survival of humans, or any life form for that matter, however, was always at the mercy of their environments, especially before people figured out how to tame the raw power of nature. Like many civilizations or cultures that vanished from apocalyptic natural disasters, the Hemudu culture ended with a catastrophic flood. Living closely to a river, the ancient Hemudu people had designed and built stilted houses that protected them from high tidal waves and occasional floods, and they could always repair and rebuild their homes once the water receded, yet nothing could save them from a deluge of biblical proportions that happened two thousand years ago, when the placid water suddenly became torrents of destruction, swallowing their houses, wiping out

THE BEGINNING OF THE STORY

their rice paddies, and forcing them to abandon their settlements.

The Hemudu site is just one of several archaeological sites in China where the remains of domesticated rice were discovered. Among them, the rice grains unearthed at the Diaotonghuan archaeological site of Jiangxi province were dated to even an earlier era—eight thousand to ten thousand years ago. Although there have been scientific debates and new hypotheses about the origin of rice domestication, most of the archaeological and genetic evidence today supports the opinion that all forms of Asian rice sprung from a single domestication event that occurred 8,200 to 13,500 years ago in China.

The Hemudu people died at least five millennia ago, yet they did not really die, as they continue to speak to us across time and space, affecting how we live and eat today. Rice is a continuous thread that connects us to them. Because of the Hemudu people and other early rice farmers, the world today has sushi and pad thai, risotto and paella, jambalaya and etouffee, and sake and Rice Krispies. Their legacies live on.

Becoming Rice

ONCE UPON A time there were no grasses on the earth. The land was mostly barren and the vegetation, if any, was dominated by trees and shrubs. Grass species emerged miraculously about 66 million years ago toward the end of the Cretaceous period, with their characteristic flimsy statures, slender leaves, hollow stems, and parallel veins. Phytoliths (siliceous plant remains) of some grass species have been found in the fossilized dinosaur dungs, implicating that the dinosaurs had eaten grasses for a brief period of time, geologically speaking, before they vanished altogether from the earth about 55 million years ago.

Today there are about twelve thousand grass species around the globe, including cereals, cattails, reeds, sugarcanes, and bamboos. The world owes its many sublime landscapes, natural or manmade, to grasses, such as rolling savannas, verdant lawns, elegant bamboo groves, and swaying marshes. But most importantly, we humans owe our very existence to many cereals in the grass family, including wheat, barley, rice, corn, sorghum, and millet.

Every cereal crop began its life as a wild grass. When wheat and barley were being domesticated from their respective wild species in the Fertile Crescent and corn from wild teosinte in Central America, a wild rice (*Oryza rufipogon*) was selected by the ancient Chinese for its starch-rich and energy-dense grains. Just like any other domesticated crop, rice has gone through many genetic changes on its several thousand years of journey from an unruly wild grass growing in the tropical swamp to a yielding crop that is currently cultivated on every continent except the Antarctic.

A wild rice plant shatters its grains. The grain falls to the ground, slips into the soil, and waits for the rain and warmth to sprout. When the right moment comes, the grain germinates and takes root. Out of hundreds of grains released from each plant, probably only a few of them ever sprout and see the light of day. After germination the seedling also needs to overcome numerous obstacles in its early life and finally grows into an adult plant, releases its seeds, and starts another cycle of life. That was how wild rice had grown and propagated for eons before its encounter with man. About eight thousand years ago a rare mutant arose.

THE STORY OF RICE

Instead of shattering and releasing its grains, the mutant kept grains on the stalk. This non-shattering trait was certainly a disaster for a plant in the wild, as its ripe seeds would never have a chance of falling into soil and sprouting, but it was a godsend to incipient farmers who had been gathering tiny grains in the mud on their hands and knees. The non-shattering grains waited wishfully to be gathered, brought home, and sowed by humans. Thus the non-shattering "freak" survived and multiplied with help from the early farmers. We now know that the change from shattering to non-shattering was caused by the mutation of a single gene, aptly called "grain-shattering gene." The lethal gene for a wild plant has become a successful gene in the plowed field.

In the wild, if all the seeds sprout quickly and simultaneously, the vulnerable seedlings might be uprooted by a gust of strong wind or scorched by a heat wave, leaving no offspring for the next generation, so the fresh seeds of a wild rice or any wild plant, for that matter, are mostly dormant and sprout in batches. As Jared Diamond said in his book *Guns, Germs, and Steel*, "many annual plants have evolved to hedge their bets by means of germination inhibitors, which make seeds initially dormant and spread out their germination over several years. So even if some seedlings are killed by a bout of bad weather, other seeds will be left to sprout later." However, this strategy used by wild plants is inconvenient and a nuisance for a farmer who wants all his seeds to sprout at the same time. The ancient people therefore either consciously or unconsciously selected seeds that germinated immediately upon sowing. As a result, the seeds of cultivated rice today are eager to sprout in response to the right combination of moisture and temperature, bursting and shooting up their seedlings simultaneously.

Today a milled grain of Asian rice looks like a white pearl, pure and simple, but the grain of wild rice is reddish, because it contains pigments called anthocyanins, which are bitter and mildly toxic to pathogens or pests. Anthocyanins evolved in wild rice to protect the grain from its natural enemies. Several thousands of years ago, farmers selected a mutant that failed to make anthocyanins, probably for the aesthetic or textural qualities of white grains. Thus the grain of cultivated rice today has shed its wine-colored coat and lost a layer of defensive armor. Consequently it has become more vulnerable to grain-infecting pathogens and grain-eating pests, such as birds and rats.

Another difference between wild rice and cultivated rice is that the kernel of wild rice has a long (~3cm) and barbed awn on its apex. An awn is a defensive structure that is akin to a spine on a porcupine or a needle on a cactus. It wards off its predators, as it is hard for an animal to swallow and eat a kernel with a long and barbed awn. In contrast, the domesticated rice has a much shortened and barbless awn on its kernel, a trait selected by farmers to facilitate

THE STORY OF RICE

harvesting, handling, and storage. Yet again, it makes grains defenseless against pests and predators, which is why a considerable portion of the grain is lost to hungry birds each year and farmers often have to erect scarecrows to guard their crops.

Furthermore, a wild rice plant is branched, prostrate, and bushy, whereas the cultivated rice has a dominant central stalk and a few side shoots—a trait selected by farmers so that they can grow more plants per unit area and therefore maximize the use of their precious land. It is analogous to the construction of high-rises to replace sprawling apartments in a crowded city such as Shanghai or New York.

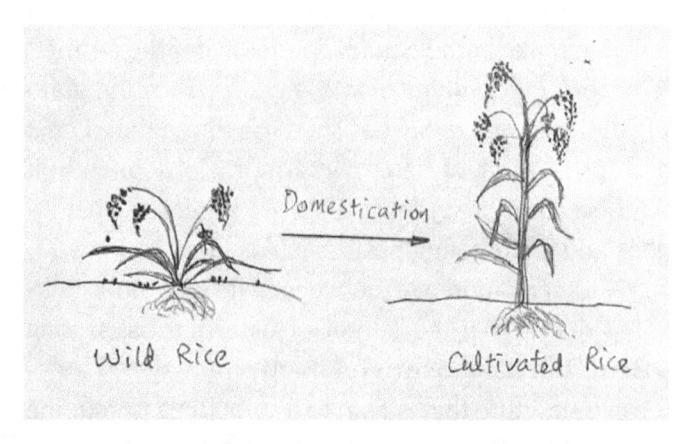

Wild rice (shattering, prostrate growth, etc.) vs. cultivated rice (non-shattering, erect growth, etc.)

In addition to these critical traits that have transformed a wild grass into a productive crop, there have been continuous improvements of other agronomic traits in rice, including yield and disease resistance. Farmers around the world have kept experimenting with rice or "trying out new rice," as expressed by the Mende people of Sierra Leone. The process of selecting desirable traits was outlined by Charles Darwin in his *On the Origin of Species*: "cultivating the best-known variety, sowing its seeds, and, then, when a slightly better variety has chanced to appear, selecting it, and so onwards." Through such relentless selection by generations of farmers around the globe and over a period of several millennia, that ancient grass has given rise to more than forty thousand varieties of rice today, with an incredible diversity for every conceivable trait.

A Tale of Two Grains

YEARS AGO A friend of mine asked me, "How do you know a house is occupied by a Chinese family?" I couldn't answer the question to her satisfaction, so she delivered the punchline of her joke: "By looking for a fifty-pound bag of rice."

The joke made me laugh as well as muse a little. Indeed, there is always a fifty-pound bag of rice in my house or probably any Chinese household, for that matter, which lasts just a few months. A large portion of it is consumed as steamed rice. The rest ends up in porridge for breakfast, often served with buns, pickles, or fermented tofu. Rice is steamed almost daily in the Hannex rice cooker sitting on the counter in my kitchen, and the house is always permeated with its lovely faint aroma.

Rice eaters in China are generally divided into two camps: short-grain eaters and long-grain eaters. Such dichotomy is due more to the geography and history than anything else. The two types of rice belong to separate subspecies. The long-grain is called indica (*O. sativa* subsp. *indica*) and the short-grain is japonica (*O. sativa* subsp. *japonica*). The indica type is cultivated mostly in tropical and subtropical areas, whereas the japonica is cold-tolerant and mostly grown in temperate regions. As a result southerners in China used to eat the long-grain exclusively because it was the only type available locally, whereas the northerners ate the short-grain for the same reason. The shape of rice grains is a reliable indicator of the texture of cooked rice. Long grains often retain their identity and remain separate after cooked, whereas short grains tend to surrender their individuality, sticking together to form soft, moist lumps. Basmati rice is a typical example of long-grain rice, a perfect type for cooking Indian-style rice, in which the "rice grains are like brothers, close yet separate, and definitely not stuck together," as said proudly by Indians, whereas sushi rice is a short-grain variety that can be easily shaped into a nigiri or norimaki because of its stickiness.

Long- vs. short-grain rice.

THE STORY OF RICE

My husband grew up in the South, eating only long-grain rice during his entire childhood and teenage years. At age of twenty, he went to graduate school in Harbin, a city in Northeastern China (commonly called Manchuria) where the best short-grain varieties were grown. He converted instantly after having his first bowl of steamed rice cooked from a local variety. He loved the way it felt in his mouth. He recalled, "I had never eaten a bowl of rice so soft before." Indeed, steamed rice from a short-grain variety gives a soft and smooth mouth feel. Porridge cooked from a short grain is even better, felt like silk or velvet in the mouth.

Whether cooked from a short or long variety, steamed rice pairs well with almost any Chinese dish or dishes. My son, who was born and raised in the States, loves eating rice more than any other food. I asked him, "Why do you like rice so much?" He replied, "Because you can make a complex meal with it." Indeed, the virtue of steamed rice is its complementary nature and balancing power. The bland rice can bring out the hidden flavor from a plain dish. It can bridge two foods that seem otherwise incompatible. It can also neutralize the strong flavor in a pungent or funky dish. The mingling of rice with meat or vegetable dishes is like a playful game that can concoct a delicious morsel out of ordinary foods.

Most Chinese agree that short-grain rice is better choice for making steamed rice, and its stickiness is ideal for eating with chopsticks. But long-grain rice is preferred to make fried rice, as long grains separate nicely so that each grain is wrapped with a thin film of oil and soy sauce and stir fried to perfection.

I learned in college that the texture of rice is really just a manifestation of starch chemistry in the grain. There are two types of starch—amylose and amylopectin. Amylose is a long, linear chain of sugar molecules, like a long rosary necklace with hundreds of sugars as its beads. Amylopectin is a branched molecule, like a tree canopy with many boughs emanating from its numerous nodes. The branches in amylopectin are easier to get tangled up once in contact with each other and therefore stick together. As a short-grain rice generally contains more amylopectin than a long-grain rice does, it tends to be stickier.

Just as any other starchy foods, once rice is chewed and swallowed, the sugars in the starch are broken off from the chain and released into the digestive tract, raising the sugar level in the bloodstream. Starch gets digested into sugars first with an enzyme called amylase in the saliva, which is why you can taste a hint of sweetness if you chew a piece of bread or a mouthful of rice long enough before you swallow. The glycemic index (GI) is commonly used to measure how fast a food raises the sugar level in blood. A food with a high GI raises blood sugar faster than a food with a low GI. As white rice is polished to such extent that there is almost nothing else left in the grain except the starch, a bowl of white rice is literally a bowl of chained sugars, definitely a food with a high GI. Yet amylopectin, the branched version of starch, is more easily broken into sugars than its linear cousin. As a result, a short-grain variety generally has a higher GI than a long-grain one, because of its higher content of amylopectin. Whole-grain rice (brown rice) has lower GI than white rice and is generally considered healthier, as brown rice contains components other than starch, such as fiber, vitamins, and minerals.

Being starch rich and easy digestibility are rice's glory and its downfall. Rice is ideal food for quick energy, fueling a hard labor or a vigorous exercise or a sport event. But if one stays idle with a full belly of rice, the blood sugar rises and remains elevated, which is why white rice has been blamed for the soaring rates of diabetes and expanding waistlines of Chinese, Indian, and other rice-eating people over the last thirty years, during which people have become more and more sedentary and did not burn every ounce of rice consumed. Harvard Medical School has constructed a Healthy Eating Pyramid, with white rice right at the peak, along with other refined grains, and suggested that rice should be consumed sparingly. Health-conscious people have already started shunning white rice and stocking up on healthier grains,

A TALE OF TWO GRAINS

such as brown rice.

I was growing up in China during the 1960s and 1970s, when white rice had a high social status among all grains and was strictly rationed. Eating a bowl of refined white rice was a sign of wealth and well-being. I ate mostly millet, sorghum, and corn, which were called coarse grains and considered poor people's food. Today rice is plentiful and common, yet I rarely eat rice to my heart's content anymore, trying painfully to trick myself to eat less—like using a smaller bowl or filling the bowl only halfway—but I am not ready to give up white rice, at least not yet, because nothing is more satisfying than a bowl of steamy and fluffy rice, especially the one cooked from a short grain.

Short and Sturdy

WHEN A FULLY formed animal is born, it has all its body parts—all of its fingers, toes, and digits. I always take delight in looking at a baby's tiny nose or ears; they are perfectly shaped miniature versions of the adult forms. A newborn will then have to grow bigger by expanding all of its body parts, more or less proportionally. But unlike an animal, a baby plant emerges from the seed with just a rudimentary root and a short stem and a couple of incipient leaves. It will constantly produce new parts—leaves and roots and branches—throughout its life. The new growth in a plant comes mostly from the special cells called meristematic cells (collectively called meristem) equivalent to the stem cells in the animal world. These special cells maintain their ability to divide and differentiate, giving rise to new cells and tissues that in turn morph into organs such as leaves and stems. Within a plant the meristem is enclosed in the tiny growth buds, barely visible to the naked eye, which include the apical bud at the apex of each stem and the axillary bud tucked in the axil area—the triangular region between the leaf and stem.

The growth of a plant starts with its main stem that is sprung out of the germinating seed. It is lengthened through the growth of the meristem within its apical bud. The stem is divided into segments by a series of nodes where leaves as well as axillary buds are attached. The segment between the two adjacent nodes is termed *internode* and does not bear any leaves or axillary buds. Such a sequence of alternating nodes and internodes is best delineated with clarity on a bamboo or sugarcane stem. The axillary bud nestled in each of the axil areas has the potential to grow into a side branch. When it does, it repeats the same growth pattern as the main shoot, producing a smaller version of it with all the parts—nodes, internodes, apical and axillary buds, and leaves. This side branch then produces its own side branches, following the same pattern. The process repeats again and again at ever smaller and smaller scales. A plant is essentially a living and growing fractal, building its body by rules of repetition and self-similarity.

Such a fractal pattern obscured by the lush green leaves in a growing season is often revealed plainly in the bleak winter, after a deciduous tree sheds its leaves. Standing under a winter tree and gazing up, you could admire its full fractal beauty, the thick trunk dividing and subdividing hundreds of times into smaller and smaller twigs from the ground up to the blue sky. It looks even more beautiful when its twigs are laden with snow or encased in ice crystals, yet please do not be deceived by the bareness and lifelessness of those winter twigs. Those dormant little buds, apical or axillary, will come back to life in spring, growing and producing more branches by the same design. There is a sense of awe in knowing that the seemingly chaotic and complex tree is a product of a simple and elegant scheme.

The infinite diversity of plants in their size and shape and branching pattern is an outcome of the varied growth of apical and axillary buds. The intricate and delicate interplay among growth buds manifests myriad forms and configurations in the plant world, such as cone-shaped conifers or bushy camellia or low-growing groundcovers, which all result from the enhanced growth in some parts of the plant and suppressed growth in others. For example, the shape of a spruce tree is a consequence of apical dominance, a botanical phenomenon whereby the growth of side branches is inhibited by a hormone released from the apical bud. The closer the axillary bud is to the terminal bud, the more its growth is inhibited. As a result, side branches gradually shorten toward the apex, resulting in a pyramid-shaped canopy—an ideal shape for Christmas trees.

Generally speaking, every plant species has its own unique form, because the overall architecture of a plant is largely controlled by intrinsic factors such as genes and hormones, yet it is also influenced and shaped by environmental factors, including light, temperature, moisture, wind, nutrient availability, or even an accident. Just like you never see two identical snowflakes in nature, you would never encounter two trees, even in the same species, with exactly the same shape. Such flexibility or plasticity in growth is an adaptive strategy used by sessile plants that cannot run away from any environmental insults or animal attacks. A plant must constantly integrate the internal information with the environmental cues in the decision whether it should stop or continue its growth or whether a bud should become dormant or active.

———∞———

The simple rule that governs the architecture of a giant tree also applies to the design of a delicate and frail grass like rice and wheat, although grasses tend to bend the rules a little. In addition to the apical and axillary buds, around the area of each node grasses have evolved a third type of meristem, called intercalary meristem, responsible for rapid lengthening of the internode during the growing season. The elongation of the main stem (or culm) in a grass is achieved initially by its apical bud, but soon is replaced by its intercalary meristems. In most grasses the first few internodes toward the base of the culm do not elongate, so the basal internodes are highly condensed and nodes are compressed and stacked together. Each node has an axillary bud that has potential to develop into a side branch, called tiller in grasses, which again is similar to the culm in its overall structure. The tiller can further produce another level of tillers and then another. As all of the tillers emerge from those compressed basal nodes at ground level, a grass often has a typical tuft-like appearance.

A grass such as rice or wheat is rarely an object of beauty or admiration. It looks simple, plain, and undignified when compared to a mighty tree. But the architecture of a rice or wheat plant holds the secret of its suitability for domestication and its yield potential as a crop. Over the history of their domestication, cereal species have shifted their architecture from profuse-tillering to low-tillering type (that is, from more tillers to fewer tillers), because of the suppression of axillary buds and enhanced apical dominance. In addition, tillers grow more erect in cultivated cereal plants compared to their wild ancestors, whose tillers grow more horizontally and sideways. Because of these architectural changes selected by farmers, more plants can be cultivated per unit of land, resulting in a dramatic increase of the yield.

One of the great successes of the Green Revolution in 1960s, which led to dramatic increases in food productivity worldwide, was the result of the change in one of the architectural traits in two major cereal

SHORT AND STURDY

crops—wheat and rice. Before the Green Revolution, the cultivated varieties of wheat or rice had tall, thin stalks. Although the taller wheat or rice plants better competed for sunlight in the field, they tended to sway and topple when hit by a heavy rain or strong wind. Especially as the grain yield increased in response to the use of fertilizer, a thin stalk could not support a heavy grain head, and it collapsed, resulting in the great loss of grain in the field. Norman Borlaug, the father of the Green Revolution and the Nobel Peace Prize laureate in 1970, tackled the problem by breeding wheat for shorter and stronger stalks that could better support larger seed heads. In 1953 his team successfully created a semi-dwarf variety of wheat by crossing a high-yielding and thin-stalked variety with a Japanese dwarf variety. In addition to its stronger stalk, the semi-dwarf variety also showed several other advantages over the traditional long-stalked varieties. First, it took less time for the plant to grow to the height needed to produce grain and therefore shortened the life cycle. Second, most of the food made by photosynthesis would be channeled into the grain instead of being invested in the stalk, further increasing the yield. This dwarf trait was later bred into many other varieties, causing a significant increase in wheat production worldwide, especially in many developing countries in the 1960s and 1970s. Following Borlaug's success, a semi-dwarf variety of rice (IR8) was developed in the early 1960s at the International

THE STORY OF RICE

Rice Research Institute (IRRI) in the Philippines using the same breeding strategy. It was created by crossing a tall, vigorous Indonesian variety with a Chinese dwarf variety. The resulting semi-dwarf rice variety was subsequently cultivated by the farmers in the Philippines, India, and other regions, boosting the yield of rice in those countries. The semi-dwarf wheat and rice were truly a game-changer. Together they jump-started the Green Revolution that transformed agriculture and changed the world forever. In combination with the use of fertilizers and good agricultural practices, these semi-dwarf varieties saved billions of lives and prevented famines and food crises in many developing countries.

IR8 is a high-yielding semi-dwarf rice variety. It was a cross of Peta, a high yield rice variety from Indonesia, and Dee-geo-woo-gen (DGWG), a dwarf variety from China. (Courtesy of the International Rice Research Institute)

SHORT AND STURDY

―――∞―――

The length of the stalk in rice or wheat is a trait controlled by a single gene, but the identity of the gene was not known until years later. Geneticists using modern molecular tools finally discovered the gene. The dwarf genes were isolated from rice and wheat respectively, and both of them were found to be involved in the biosynthesis of gibberellins (or GAs)—a group of multitasking hormones that regulate many physiological processes, including the elongation of the stem. A mutation of the gene rendered the plant short and sturdy, just the right trait Norman Borlaug and other breeders sought in the 1950s and 1960s.

Shape really matters, because it affects many physiological processes in a plant, such as photosynthesis and allocation of sugar into grains. So far, many genes controlling the architectural traits of rice and other cereal crops have been identified and cloned, including the genes that control the degree of tillering, the angle between tillers with the main stem, the orientation of leaves on stem (erect, oblique, or drooping), and the shape of seed heads, all of which were found to be closely connected with the grain yield. In some cases, a single gene can act as a master gene, controlling several architectural traits simultaneously. For example, a gene called IPA1, which stands for ideal plant architecture, has been found to regulate multiple architectural traits concurrently, including the degree of tillering, the length of stalks, and the shape of seed heads. However, unlike the dwarf trait, many of those architectural traits in rice are complex traits, meaning they are regulated by multiple interacting genes and therefore cannot be easily altered or manipulated by traditional breeding.

Today rice researchers are not trying to find a magical gene that would ignite another green revolution, but they are assembling a set of genes and using them to design rice with an ideal form, a form capable of absorbing sunlight efficiently and allocating sugar preferentially into grains. Such "designer rice" would have blades with perfect angles for light interception and seed heads with an ideal pattern for maximum food storage in the field. It is anticipated that in the future rice—or any crops, for that matter—can be designed and reinvented, just like an industrial product, to have a form intimately connected to its functions. In such a plant all the genetic components are integrated and their relationships are optimized so that they will work together to improve the yield.

Hybrid Rice

UNLIKE MOST OF the animals we are familiar with, plants are generally hermaphrodites or bisexuals. A typical flower—the sexual organ of a plant—consists of both male and female parts, in which sperms are produced in the miniscule pollen grain and eggs are hidden in the depth of the flower. Once an egg is fertilized by a sperm, the resulting zygote will start its journey to become an embryo and then an adult plant, an odyssey as fascinating as its animal counterpart.

Most of the wild plants are self-incompatible hermaphrodites. It means the plant is incapable of fertilizing itself, and it has to receive pollen from a different plant. Although such self-incompatibility ensures the diversity of offspring and provides them with selective advantages, it was an undesirable trait to an ancient farmer, because when a rare mutant with a favorable characteristic emerged, the farmer would want to preserve the mutant. If the mutant could not self-fertilize and had to produce its offspring with a "normal" plant, the good mutation would be diluted or lost. In an effort to preserve a desirable mutant, ancient farmers probably unconsciously selected the hermaphrodites that had lost their self-incompatibility and become able to self-fertilize, so the mutant would automatically be preserved and the favorable trait passed on to its offspring, which is why many of our domesticated fruit trees and cereal crops are self-pollinating and self-fertilizing.

On the other hand, the crossing between two varieties often results in a vigorous hybrid, a phenomenon called heterosis (or hybrid vigor). A hybrid often exhibits superior traits, presumably because of the suppression of undesirable recessive traits from one parent by desirable dominant traits in the other. As a result, a hybrid unites the good traits from both parents.

Heterosis has long been exploited to breed animals. For example, Black Baldy is a hybrid produced by crossing Hereford cattle with a solid-black breed, for its good mothering instinct and black skin that protects the cattle from the scorching sun. In another example a broiler is a hybrid poultry produced for its breast meat. Sometimes a hybrid can be created by crossing two species. For example, a mule is the progeny of a female horse and a male donkey and famous for its patience, hardiness, intelligence, and longevity.

In the plant world, hybridization seems to be a rule rather than an exception. Charles Darwin, best known for his theory of evolution, was keenly aware of the phenomenon of heterosis in plants. In his book *Variation of Animals and Plants under Domestication*, he cited many instances of heterosis, from melons and apples to cabbage and wheat. Even though he admitted he did not understand what was behind this biological phenomenon, Darwin suggested that hybridizing could be used to create new varieties with superior traits.

Luther Burbank (March 7, 1849 – April 11, 1926), an avid American horticulturist and breeder in his time, was inspired by reading Charles Darwin's book and set out to produce hybrids. During his life, Burbank made at least thousands of crosses—peaches with plums, apples with quinces, potatoes with tomatoes, and many others. He produced some superb hybrids that have lasted to this day, including Shasta daisy, Santa Rosa plum, and white blackberry. Yet his relentless and single-minded crossing and hybridizing through trial and error, without guidance from any solid scientific theory or principle, yielded contradictory results and some useless hybrids, which were promptly burned and destroyed. Burbank's work was certainly useful and practical but had limited scientific value and offered little help for other breeders, especially cereal breeders who were trying to increase food production and alleviate the widespread hunger at the time.

THE STORY OF RICE

It had been very difficult to develop a hybrid from self-pollinating cereal crops. To do so a breeder has to remove anthers, the pollen-bearing blobs, from a plant, preventing the plant from self-pollination, and then introduce pollen to it from another plant. Once the seeds from such a crossing are collected and sown, they grow into hybrid plants—often called F1 hybrids—that often exhibit hybrid vigor. If you gather and sow the seeds from a F1 hybrid, they will sprout and grow into a new generation of plants called F2 hybrids, but these F2 hybrids bear little resemblance to F1 hybrids, because of the genetic reshuffling, and lose most of their superior qualities. Hybrid seeds therefore have to be generated afresh every year through crossing two parental strains.

In the 1920s corn (or maize) was the first cereal crop that was successfully used to create hybrids. The early success of corn hybrids was largely because of the peculiar sex of corn. Unlike rice and wheat, the male and female flowers in corn are borne on separate parts of the stalk. Its male inflorescence, commonly called tassel, forms at the apex of the stalk like a bouquet, and produces hundreds of anthers. About a meter below the tassel, a female inflorescence (or ear) nestles at the crotch of a leaf, where hundreds of miniscule female flowers are arranged in straight columns along a tiny cob wrapped tightly in the layers of husk leaves. Each female flower has the potential to become a kernel if pollinated. Over the course of several summer days, the tassel showers the corn field with a deluge of microscopic pollen grains—about 14 to 18 million grains per plant. Simultaneously, each of the female flowers in the cob sends out, through the tip of the husk, a long and sticky strand of silk (botanically it is called *style*) to guide pollen grain into its concealed ovule. Once the egg in each ovule is fertilized by the sperm from the pollen grain, the female inflorescence grows and swells, manifesting itself a few months later as a hefty ear swaddled in a green husk and draped with a tuft of reddish silk at its tip. Once layers of husk are peeled off, inside are the neat rows of yellow kernels we all love to nibble.

The strange sex of corn seems to be designed to welcome human intervention in its sexual life. Breeders can easily emasculate a corn plant by removing its tassel on the top, a process called de-tasseling, and then introduce the pollen of another variety. The hybrid is expected to combine the good traits from both strains carefully selected for this purpose. Before the arrival of the detasseling machine, seed companies and breeders had for years hired an army of detasselers each summer to remove the tassels in their cornfields. Detasseling has been a summer job for middle and high schoolers for many years in the Corn Belt and still is today.

THE STORY OF RICE

Rice is an entirely different story. In rice both male and female parts are borne in the same flower. Removing the male organ from every tiny floret in a rice field is an impossible task, so some rice scientists had previously thought that rice hybridization could not be achieved in an agricultural scale. Some researchers even doubted the existence of heterosis in rice. In the late 1950s and early 1960s, Yuan Longping, a young rice researcher in China, kept thinking about how to improve rice yields. On a hot summer day in 1960, a giant rice plant caught his attention while he was working in a paddy. The plant stood out in the field like "a crane stands out among chickens," he said. He noticed its larger-than-usual seed head and counted 230 grains—more than the average, then. He thought he had found a high-yielding mutant. Naturally he saved all the seeds from the plant and sowed them, yet the next generation was a mix of plants with inferior characteristics, nothing like the parent. Initially it was a huge disappointment, but it was also a eureka moment, because he realized that the plant was a natural hybrid produced by hybridization—an exception to the rule for self-pollinating rice. He collected data and confirmed his speculation. The discovery of this natural hybrid inspired him to focus his work on developing hybrid rice.

Yuan came up with an idea and designed a scheme of producing hybrid rice. He proposed to develop a male-sterile strain of rice and then pollinate it with another strain to create a hybrid. He prophesied that a mutated rice plant with male sterility might exist somewhere, and once found, the male-sterility trait could be bred into a cultivated variety to create a male-sterile line. Based on his theory, he and his assistants started a two-year long search in the field for the hypothetical male-sterile mutant. During the months of June and July, Yuan and his assistants examined tiny florets with a magnifier, one flower a time, standing under a scorching sun and among thongs of insects. After scrutinizing more than 12,000 plants for two years (1964 and 1965), they identified six precious male-sterile plants. When I heard this story, I conjured up the images of Gregor Mendel counting more than 300,000 peas to see the genetic pattern and Marie Curie and her husband boiling tons of pitchblende to extract a tiny vial of radium. Those great scientists moved scientific mountains with their hands, shovel by shovel.

In 1966 Yuan published a seminal paper to delineate his plan and started his breeding program right away. Those first plants were not the ideal he had hoped for, as the male-sterile lines developed from them still occasionally produced viable pollen and underwent self-pollination. As a result, the hybrid seeds were mixed with many seeds resulting from self-pollination. He then came up another idea—finding a male-sterile plant from wild rice and using the trait to generate a male sterile line. He started another epic search, but he could not find one until several years later. One day in 1971, Yuan's assistants spotted one plant among a swath of wild rice in a ditch in Sanya, of Hainan Island in China. The plant stood with three panicles of flowers, all of which had aborted male organs. They pulled the plant out of the swamp carefully and wrapped it gingerly with a paper. Covered with mud and dripping with water, his assistants brought the plant back to the lab. They nurtured the mutant like a baby, hand-pollinating all of its sixty-five florets. In the end, they harvested five golden grains, in which the dream of creating a male-sterile line glimmered hopefully.

It is a story beyond "finding a needle in a haystack." It is a story of faith, perseverance, and momentous encounter at the end of the earth—literally, because Sanya was considered by ancient Chinese as the end of the earth because of its remoteness from the mainland. After thousands and thousands of crosses, the male sterility trait was successfully bred into a cultivated variety and a male-sterile line was established. In 1974 the first hybrid rice was created using this line, and its yield was about 15 to 20 percent greater than those of traditional varieties. Many hybrids followed, with high yield and other desirable agricultural traits.

Today hybrid rice varieties developed by Yuan and other scientists account for more than 50 percent of rice grown in China. They are also grown in many other rice-cultivating countries, including India, the Philippines, Vietnam, Indonesia, Myanmar, Bangladesh, Sri Lanka, Brazil, and the USA, because

HYBRID RICE

of their high yield or other superior agronomic traits. Hybrid rice has helped combat world hunger and food crises since its invention in 1970s. Yuan is hailed as the father of hybrid rice and won numerous prestigious awards, including the Wolf Prize in Agriculture and the World Food Prize in 2004.

Yuan Longping in the rice field.

THE STORY OF RICE

---ꝏ---

An uncanny coincidence happened to me while I was writing this chapter. Yesterday I completed the chapter with an ending: "Yuan, in his nineties now, is still leading the way. He and his team are currently in the process of developing high-yielding rice he dubbed Super Rice and salt-resistance varieties. He would never stop." I then heard the news of his passing at the age of ninety. All of China mourned his death. Tens of thousands of people in his hometown went out into the streets and bade farewell to Yuan when his hearse passed by. It was a rainy day, as if the sky were also crying along with the mourning crowds.

Never before in China had we seen such an outpouring of grief over the death of a scientist. Yuan never discovered a new law of genetics or made a groundbreaking invention, yet to the Chinese, nobody was more lovable and dearer than a person who helped fill up their rice bowls to the brim and who had saved many people's lives from starvation in the 1970s and 1980s.

In Chinese culture Yuan was the epitome of what a scientist or a good person should be. He was quiet, humble, diligent, and morally intact. His success came from his many years of ponderings and dogged labors. He had gone to the field almost daily until he was hospitalized two months prior to his passing. His intimate knowledge of rice set him apart from other researchers, so he could see things unseen by others and discern an unusual plant that had escaped the notice of his colleagues. He also shared his innovations unselfishly and distributed his precious male-sterile plants to his peers, turning his individual brilliant ideas to a nationwide breeding effort.

Part of Yuan's charm was that he looked like a farmer. His weather-beaten and sun-drenched skin were an eloquent testimony to his almost seventy years of sweat and toil in the rice field, yet he was also a strong swimmer and an amateur musician. He enjoyed playing violin, holding the bow and plucking the strings with hands roughed from years of the grueling work of planting and crossing. His music and science intermingled, giving him the aura of a renaissance man. He said he found harmony and serenity while playing violin, echoing what Einstein had said.

Yet Yuan was a man of his time. By the end of 1980s, geneticists had begun to use molecular techniques to improve crops by transferring genetic material directly. The science and art of plant genetic engineering had advanced far beyond anything Yuan ever attempted and achieved in the 1970s and 1980s. Even a male-sterile plant he had searched for years in the field could be created in the lab with molecular tools. In his later years he tried to stay relevant, proposing fresh ideas and schemes to create new varieties of hybrid rice. He and his teams used molecular technology to help his breeding programs. In one case his team used a gene-editing tool to knock out a gene for absorption of cadmium (a toxic metal) in rice and bred a variety with a low-cadmium content even when grown in cadmium-contaminated soil.

While borrowing molecular tools for his breeding program, however, Yuan made some negative comments about the creation and safety of genetically modified organisms (GMOs), and therefore he might have played a significant role in promoting anti-GMO sentiment among Chinese people, largely due to his fame and reputation. Some scientists think that Yuan's attitude toward GMOs actually hindered and delayed the progress of plant technology in China. Still, Yuan deserves a special place in the history of rice, and his hybrid rice is one of the most important chapters in the story of rice.

Golden Rice

PLANTS GRACE OUR landscapes and decorate our gardens with their thousands of colors and hues, which are actually a visual language used by plants to communicate with animate beings. A bright yellow flower may lure a honeybee to its concealed nectar. While buzzing from one flower to another, the bee unknowingly transports pollen for the plant. A flaming red cherry fruit may attract a bird who would gulp the sweet meat and defecate or spit out the hard pit in a faraway place, incidentally helping the cherry tree disperse the seed. Plants and animals have communicated this way for eons, ever since the emergence of flowering plants on the earth more than 125 million years ago.

Although the colors in a plant seem intriguing and tantalizing, they are really just the manifestation of a plant's chemistry. Plants are capable of producing a variety of pigments, and each absorbs sunlight at particular wavelengths and reflects at the others, resulting in a characteristic color. The yellows, oranges, and reds are mostly emitted by carotenoids, a group of pigments packed in miniature structures called plastids within a plant cell. The combinations among various carotenoids give rise to every possible shade and tone of yellow or orange or red, as often seen in flower petals, autumn leaves, and fruits. In the plant world these are often the colors of ripeness that enable animals to distinguish mature from immature fruits. A field dotted with red tomatoes or a grove of trees laden with oranges or a patch of land blanketed with yellow squash all beckon animals to come have a feast.

Some of the carotenoids in plants, such as α-carotene and β-carotene, happen to be the essential ingredients in the human diet, as they are needed to make vitamin A, a key micronutrient for us to maintain a strong immune system and healthy eyes. Fortunately we are bestowed with a variety of foods that are rich in carotenes, including several staple foods such as corn, sweet potatoes, and yams, as hinted by their yellow or orange hues. Many fruits and vegetables also contain a high content of carotenes, exemplified by cantaloupes and carrots. People who eat these foods regularly automatically receive enough carotenes, which in turn are converted into vitamin A in the body.

Rice, however, a staple food for many people around the globe, is devoid of carotenes, which poses a major health problem in some parts of world, especially in Southeast Asia and Africa, where people eat mostly rice and do not have much access to carotene-rich foods. As a result people in those areas often suffer from vitamin A deficiency (VAD), the leading cause of childhood blindness. A deficiency also causes a weak immune system in children, increasing their risk of dying from even common infectious diseases such as diarrhea.

In an effort to combat the problem with VAD in those areas, rice researchers had pondered about breeding a rice variety containing β-carotene. Peter Jennings, a rice breeder at IRRI in the Philippines, was one of them. While breeding rice for higher yield, Jennings also kept thinking about the nutritional quality of rice. He hoped to find a wild rice with yellow endosperm (like the one in corn or millet) so he could breed the trait into the rice varieties commonly grown in Southeastern Asia and Africa, but he never found one.

While attending a meeting at IRRI in 1984, Jennings and several other rice breeders gathered to discuss how to improve the quality of rice using the emerging plant genetic engineering technology. According to a legend, an idea was conceived at the guesthouse at IRRI over a couple of beers. Jennings proposed to create a rice cultivar with yellow endosperm by borrowing the "yellow genes" for making β-carotene from other organisms and putting them into rice. Gary Toenniessen, who was in charge of the biotechnology program at the Rockefeller Foundation at the time, liked the idea and believed it could potentially save the lives of a hundred million children who were at risk of suffering from VAD each year in countries where rice was the only staple food.

Toenniessen then funded several research labs around the world to create a rice variety with β-carotene. Scientists isolated two "yellow genes" from daffodils and a soil bacterium, both of which were required for synthesizing β-carotene. The two genes were "blasted" literally into rice with a gene gun. Other molecular tricks were also devised by scientists to make sure that β-carotene was made only in the endosperm, the way it was made in corn and millet. It took at least eight years for a large group of scientists led by Ingo Potrykus and Peter Beyer to produce the first grain of yellow rice—named Golden Rice. The first-generation Golden Rice did not produce sufficient β-carotene that could meet the dietary need of people who were suffering from VAD, though. A team of scientists at Syngenta, a biotech company, did more molecular tinkering, creating a second generation of Golden Rice, aptly named Golden Rice 2, that produced twenty-three times more β-carotene than the first-generation Golden Rice. The following studies confirmed that β-carotene in Golden Rice 2 was safe to eat and could be effectively converted into vitamin A in humans. A single bowl of this rice can supply more than half of a child's daily vitamin A requirement.

By the 2010s, other types of genetically modified plants, such as soybean, cotton, papaya, and potato, were growing on farmlands in many countries. Golden Rice still hadn't been approved yet, much less saved children who were at risk of VAD. The scientists and supporters of Gold Rice met tremendous resistance when testing and planting Golden Rice. Greenpeace led the opposition to growing Golden Rice, or any GMOs, for that matter. In 2013 protesters uprooted and destroyed an experimental plot of Golden Rice being developed at IRRI.

In 2008, in the midst of the global debates and controversies about Golden Rice or other GMOs, an incident involving Golden Rice caused an unprecedented public and media uproar in China. It revolved around a trial conducted on Chinese children who were fed Golden Rice. It was a study done by Tufts University in the United States with its Chinese collaborators. This trial was not about the safety of Golden Rice, which was already proven and documented by the earlier studies. It was an efficacy study to determine if the β-carotene in Golden Rice worked as well as β-carotene supplements. The results showed that β-carotene in Golden Rice indeed worked equally well. However, the researchers didn't receive all the approvals needed for the study, and they deliberately hid the fact that the rice was genetically modified. Many Chinese were furious at the idea of anyone using their children as guinea pigs and feeding them with genetically tinkered rice. The Chinese government reacted quickly to the public outcry this time, punished the three Chinese co-authors of the study, and offered monetary compensation in the amount of ¥80,000 (more than $ 10,000) to each of the families whose child was fed Golden Rice. Following the response of the Chinese government to the incident, Tufts University announced that one of its researchers had broken ethical rules while carrying out the study, even though the scientific conclusions of the study remained valid.

Just before the incident, many Chinese were already deeply worried by the earlier report about the pollutant-tainted rice grown in some parts of China where soil had been contaminated with cadmium and other toxic heavy metals from industrial wastes and sewage disposals. The Golden Rice incident made them even more concerned about the safety of their traditional food. The brilliant yellow grains of Golden Rice might look appealing to those who ate saffron rice, but Chinese were so used to snow-white rice that the golden grains looked strange and unnatural to them.

The negative sentiment toward the Golden Rice incident quickly spilled over to other GMOs. GMOs or non-GMOs became the issue debated on TV, newspapers, and blogs. In the summer of 2013, I signed up for WeChat, a smartphone app developed by Tencent in China, and soon I started receiving anti-GMO messages, often accompanied with appalling pictures, such as a lab rat with bulging tumors as big as a ping-pong ball after being fed genetically modified food. Such shocking images were circulated and recirculated among 300 million Chinese users with a touch of the fingertip. Rumors spread like wildfire. Among many outrageous claims, the one that really touched a sour spot in Chinese people was that genetically modified food caused a decline in fertility among lab rats and pigs, citing an unpublished work. A rumor even claimed that the quality of sperm in male college students in China was reduced after they ate genetically modified food. As getting married and having a son to carry on the family line was still deeply rooted in Chinese culture, especially in the rural areas, many people believed that GMOs would bring the worst curse upon their families and children. Filled with such fear, more and more Chinese people became strong opponents of Golden Rice and GMOs in general.

Even though Golden Rice has been demonized by organizations such as Greenpeace and people who oppose GMOs, it has been supported by the Bill and Melinda Gates Foundation, blessed by Pope Francis, and endowed by 107 Nobel laureates who signed a letter urging Greenpeace and its supporters to abandon their campaign against Golden Rice. In May 2018 the U.S. Food and Drug Administration finally approved the use of Golden Rice for human consumption, joining Australia, Canada, and New Zealand, which issued their approvals earlier, in 2018. In 2021 the Philippines becomes the first country to approve Golden Rice for planting. Finally Golden Rice reached poor countries and helped those at risk of VAD, as intended by its inventors and advocates.

Golden Rice has come a long way since the day when the idea was conceived almost forty years ago, yet nobody is sure that Golden Rice is the best or only solution that prevents or treats VAD, because ultimately there are other options or even better ones to combat the problem of VAD. Instead of relying on a single staple crop, time and money could be spent on helping people grow crops that are naturally rich in β-carotene, such as sweet potatoes, carrots, or leafy vegetables—a much healthier and more sustainable approach of fighting VAD.

Golden Rice, after all, represents more than just another genetically modified food. It symbolizes a new concept and idea. To some people it was a bad idea that could open Pandora's box and unleash all kinds of Frankenfood in the future. To other people it is an innovation that could lead to new opportunities of improving the nutritional quality of food. If scientists could put β-carotene into a grain of rice, they could definitely create a food containing any other nutrients, supplements, or even medicines such as iron, folic acid, minerals, vitamins, and vaccines. In the future, as some scientists imagined, when you have a meal, you could also be taking your supplements or pills. The line between food and medicine would be blurred. Is it a sweet dream or a scary nightmare? It all depends on our individual attitudes and ideology, which are shaped by our experiences, personal values and beliefs, and hundreds of other things.

Rice is the staple food for nearly one-third to one half of world's population, despite its poor nutritional quality. It is therefore impossible to be replaced with any other foods. Rice is here to stay, and it will accompany us, our children, and our children's children far into the future. It is imperative for scientists to develop rice varieties with high nutritional values one way or another, and genetic engineering technology used for creating Golden Rice is just one of the approaches that can be used by scientists to create nutrient-enriched rice varieties.

Zongzi

ON A HOT summer day in 2015, our family of four arrived at a hotel in Hangzhou after traveling for a whole week across China. Much to our surprise, we received a big box as a welcome gift at the front desk. The box was shaped like a giant zongzi (a tamale-like food) and decorated with Chinese paintings and calligraphy. My husband and I realized suddenly that it was the Double Fifth Day—the fifth day of the fifth month in Chinese lunar calendar. It is the third of the mostly celebrated traditions by Chinese, following Chinese New Year and Mid-Autumn Festival.

We opened the box in our room and saw twelve pieces of zongzi snugged in the box. What a thoughtful gift for holiday travelers, I thought. The food evoked a sense of nostalgia in my husband and me and we told our two children the story of zongzi.

Zongzi is a traditional Chinese food made of glutinous rice stuffed with flavorful fillings and wrapped in a bamboo leaf. It is a simple and delicious food deeply rooted in Chinese history and intertwined with myths and legends. The origin of zongzi is closely connected with Qu Yuan (340 – 278 BC), a famous Chinese poet and statesman who lived during the Warring States period of ancient China (475 to 221 BC). Many people regard Qu Yuan as iconic a figure as Confucius has become in Chinese history. If Confucius was comparable to Socrates and Plato in the West, Qu Yuan would be the Eastern equivalent of Shakespeare or Dante.

During Qu Yuan's time—a time of constant political and military upheaval—China was divided among the seven Warring States—Qin, Han, Wei, Zhao, Qi, Chu, and Yan, whose allegiances and rivalries shifted constantly. Qu Yuan served as an official in the State of Chu under the ruling of King Huai of Chu (328 – 299 BC). In the midst of the wars among seven states, Qu Yuan opposed King Huai's policies toward other states and advocated that Chu and Qi should join their forces to fight together against Qin's expansion. To punish Qu Yuan's opposition, King Huai exiled him to a region near the Miluo River in today's Hunan Province, where Qu wrote several classics, including *Tianwen* (translated to *Quest for Heavenly Truth*) for which China's first Mars exploration mission in 2020 was named. Learning that Qin's army had taken Ying, the capital of Chu, Qu Yuan wrote a lengthy poem titled *Lament for Ying*. On the fifth day of the fifth lunar month of that year (278 BC), he held a big rock

THE STORY OF RICE

and plunged himself into the Miluo River.

According to a popular legend, villagers rowed their boats to the middle of the river, trying desperately to save Qu Yuan, but failed. They then threw rice into the river, both to lure fish away from Qu Yuan's body and to serve as a food offering to his departed spirit. One day an old man met Qu Yuan in his dream and asked Qu, "Did you eat the rice we threw into the river?" Qu Yuan replied, "No, it was eaten by fish, turtles, and crabs." The old man asked, "How can we prevent them from eating it?" Qu Yuan said, "You can wrap the rice in a leaf." That story was the earliest version of zongzi, just a package of plain rice. Over the years people in different parts of China have jazzed up zongzi with all kinds of fillings—meat, beans, or fruits. Making and eating zongzi on the fifth day of the fifth lunar month, the date when Qu Yuan died, has become a cultural event of honoring Qu Yuan and remembering that tumultuous time of Chinese history. That date has been aptly named Double Fifth Day and set aside as an official holiday in China.

Dragon boat racing has also become a tradition on Double Fifth Day, reenacting the scene of villagers rowing boats and searching for Qu Yuan in the Miluo River 2,000 years ago. Double Fifth Day is therefore also called Dragon Boat Festival. Unlike the fire-breathing Western dragons, Chinese dragons are the rulers of running water—rivers, seas, and waterfalls. A dragon boat is built in the image of a dragon, long

ZONGZI

and slender, with upturned ends carved into a dragon head and tail respectively. The helmsman, wearing a red turban, sits on the prow, waving two flags to direct the boat. Two men sit in the middle, beating a drum and gong to set the tempo. Twelve to eighteen oarsmen holding short paddles, row the boat in unison. Once the boat is launched, those oarsmen become one, thrusting their oars into the river to the rhythm of the drumbeats. From a distance those boats look like arrows shooting across the river, leaving long white tails in their wake.

Dragon boat racing, a 2,000-year tradition and as old as the Olympic Games in Ancient Greece, has spread all over the world and evolved into an international sporting event, beginning in 1976 in Hong Kang. Today it is one of the popular team water sports, with millions of participants in nearly one hundred countries around the globe. Dragon boats have shown the whole world their glamorousness and speed when they ferried torchbearers across a river or creek during the ceremonious torch relay of several Olympics.

———⧖———

On a chilly January morning, I was listening to National Public Radio on my way to work and heard an incredible story. I was touched by it and had to finish listening to it in the car when I arrived at work. The story began with a dragon boat racing event in

THE STORY OF RICE

Florence, Italy, and all the competing athletes, 3,500 of them, were breast cancer survivors. As the story continued, I learned that the sporting event started in 1996 with twenty-four Canadian breast-cancer patients who challenged the conventional wisdom that a breast-cancer patient should avoid any repetitive upper body motion, even something like cutting vegetables and raking leaves, to prevent lymphedema, a condition often associated with breast cancer. Under the guidance and supervision of a physician specialized in sports medicine, they formed a team called Abreast in a Boat and started their training in dragon boat racing. They chose the dragon boat over other types of canoes because of its stability in the water. Six months later, they proved that racing didn't cause any development of lymphedema, a finding published in the *Canadian Medical Association Journal* in 1998. They continued, spreading the news and encouraging other breast-cancer patients to paddle on a dragon boat. Over the two decades the sport has evolved into an international sporting event with participants from all over the world.

Later that day I found an online version of the story and saw the beautiful pictures of the athletes rowing and smiling in vibrant pink shirts. Of the twenty-four original members of Abreast in a Boat team, twenty-two were still alive and well. Dr. Susan Harris, one of the original team members and a physical therapist and professor at the University of British Columbia, published her own article titled "We are all in the Same Boat: a review of the benefits of dragon boat racing for women living with breast cancer" in a medical journal. Jane Frost, another member of the original team, founded and became the first president of the International Breast Cancer Paddlers Commission. I was amazed by the way the ancient Chinese dragon boat found its way and its new advocates in Canada and by the unlikely juxtaposition of dragon boats with breast cancer. The 2,000-year-old dragon boat has been given a fresh life and new purpose, making an impact on the lives of breast-cancer patients.

Making zongzi is a two- or three-day affair. I remember my grandma always starting to prepare ingredients a couple days before the Double Fifth Day, soaking glutinous rice and bamboo leaves in water to soften them and marinating the meat in soy sauce. The commonly used fillings in our region then were red bean paste, Chinese dates, and pork. My grandma often made two kinds—a salty one and a sweet one—to accommodate differing palates in our big family.

Wrapping zongzi was a family event. With a big pot of snow-white rice, a heap of lustrous-green bamboo leaves, and small bowls of fillings crammed on the table, we sat around wrapping, chatting, or just watching. It was always fun for small kids to watch adults making zongzis. I still remember vividly my grandma shaping the leaf into a funnel with a swift twist, adding a spoonful of rice into it, throwing in a piece of filling—either a sliver of meat or a plump date—and filling the funnel to its brim with more rice. She gave the rice a good pat and covered it with the remaining leaf. With one hand holding the rice wrapped in a leaf, she used her free hand to tighten it with a twine, with one end anchored between her teeth. In a minute a zongzi with a perfect geometric shape—either cone or pyramid—was wrapped up. In no time the big pot was piled with finished products. Watching my grandma wrapping zongzi swiftly and effortlessly, I couldn't resist trying with my own hands, yet my small, clumsy hands always yielded a zongzi with an odd shape, white grains spilling out all over.

It is not easy to wrap a handful of loose grains into a narrow saber-shaped bamboo leaf. It is a learned skill and needs lots of practice. The bamboo leaf is not just simply a wrapper. It has a unique fragrance, earthy and refreshing. Bamboo leaves, boiled or steamed with rice and whatever the filling inside, permeated the whole apartment with a distinct scent, and the aroma could linger for hours or even days.

The rice used for making zongzi is glutinous rice—a "genetic freak" in the rice family that produces nearly 100 percent amylopectin—a branched form of starch. Those amylopectin molecules are loosened up by the heat and moisture during cooking and get tangled with each other like a big mess of wires; as a result a fistful of loose rice grains turns into a soft lump. I was amazed by such a magic transformation when I was a child.

Zongzis made from glutinous rice.

Eating zongzi is delightful and playful for a child. Once the string is untied and the green leaf unwrapped, a perfectly shaped and pure-white zongzi sits in its rumpled green garment, steamy and glistening. I remember plunging my chopsticks into it, searching eagerly for the filling until the hidden jewel was revealed. I often saved the filling, the most flavorful part of a zongzi, for last. To an undeveloped child's palate, the filling is more desirable than the bland rice.

My grandma always cleaned and saved the

bamboo leaves and twine for next year. Nothing was wasted then. Saving and recycling were a way of life back in the 1960s and 1970s. Zongzi was always on the table on Double Fifth Day each year. There were times when money was tight and food was scarce, yet my grandma still made them anyway, plain ones without any filling inside, which we ate happily with a sprinkle of sugar.

While I was watching *Ratatouille*, a computer-animated movie, my eyes welled up when Ego, the food critic, reminisced how his mother's ratatouille cheered him up after he had a bad day at school. Everyone has his or her ratatouille. For me it is zongzi. Just like the simple vegetable dish that takes Ego to his childhood and his mother's cooking, zongzi is a thread that connects me to my long-gone grandma and the place where I grew up.

Although many traditional foods are disappearing today, zongzi still stays and keeps evolving, simply because the 2,000 years of history and culture are wrapped craftily into this food, along with the traditional ingredients.

Guilin Rice Noodles

GUILIN RICE NOODLES are a local specialty that has become a symbol of Guilin—a southern city in China. According to a legend, the invention of Guilin rice noodles was associated with an ancient bloody war fought in the city about 2,300 years ago during China's Qin dynasty. In 214 BC, Qin Shi-Huang (literally the first emperor of China) sent his army south to conquer the territory of the southern tribes after his successful unification of China. Qin's army was made up of 500,000 strong and well-trained warriors equipped with the most advanced weapons at the time. One could get a glimpse of the sophistication and grandeur of Qin's army from the Terracotta Army buried with the emperor in Xian (the capital of Qin) when he died in 210 BC. The Terracotta Army was constructed to accompany the emperor to his underground kingdom and protect him in his afterlife. The excavated army is composed of more than 8,000 life-sized and vividly sculpted warriors, 130 chariots with 520 horses, and 150 cavalry horses. Those warriors look strong and fearsome, standing in a military formation with their weapons as if they are marching toward war and ready for a fight.

The real soldiers who fought the war in Guilin, however, came from the northwest of China. To these soldiers the south was a foreign and unknown world to them. They were lost in the jungle terrain and succumbed to various tropical diseases. Furthermore, those young men grew up eating wheat noodles and other flour-based foods, and rice didn't agree with their stomachs. As a result the army was initially defeated by the southern tribes.

At the saying goes, necessity is the mother of invention. The military chefs created rice noodles in the midst of the war. They ground rice grains into flour, kneaded the flour into dough, and shaped the dough into noodles. To help soldiers combat infectious diseases, military doctors also added herbs and Chinese medicines in the noodle soup; thus the first bowl of Guilin rice noodles was born in a military camp more than 2,000 years ago. To the soldiers, rice noodles echoed the taste and smell of the food they ate at home and lessened their hunger and fatigue. The herbs and medicines blended in with the soup also effectively protected the soldiers from tropical diseases and other discomforts. It is not an exaggeration to say that the rice noodle helped Qin's army win the war. Indeed, nourishing food is very essential for soldiers

GUILIN RICE NOODLES

in combat, as said vividly by Napoleon or Frederick the Great: an army marches on its stomach.

———⚭———

Whether strolling along the bustling streets or weaving through the narrow alleys during our family trip to Guilin, we always smelled the air permeated with the aroma of rice noodles—the permanent smell of the city. We saw people bending over steamy bowls and slurping up the long strands noisily at noodle shops and food stalls everywhere. Like Goro, the truck driver and a noodle fanatic in *Tamporo* (a hilarious Japanese comedy about noodles), we wandered around in pursuit of a perfect bowl of noodles in town. A bowl of Guilin rice noodles looks very much like a bowl of ramen—long ribbons sinuating in a steamy, clear broth; pearls of oil glittering on the surface; green bits of scallion or other vegetables drifting in the soup; and thin slices of pork belly with honeycombed golden skin floating on the top. But when I slurped up the noodles, they were unique; the noodles made from rice were tender and springy, squiggling in my mouth. The pork belly meat tasted even better; it was fried to perfection—tender meat fringed with a crispy layer of skin. But what held all the ingredients together was the broth. Beneath its clear appearance, the broth was a rich union of flavors and tastes, delicious beyond description.

THE STORY OF RICE

To make a perfect broth, a concoction of meats, vegetables, fruits, and spices are simmered together for hours or even days in a big pot. The low heat extracts the essence of each ingredient, building layers of rich flavor and taste. The noodle soup is truly a gestalt—the whole is greater than the sum of the parts. There are many ways of making a noodle broth. The ingredients each restaurant uses and the way those ingredients are combined and simmered in the pot are closely guarded secrets and passed down through generations. A variety of recipes and cooking techniques has resulted in hundreds of noodle shops in Guilin alone, each of which has a style of its own.

A few specific ingredients, however, give a characteristic taste and flavor to Guilin rice noodles. Dozens of Chinese traditional herbs and spices are often cooked into the broth. Their antimicrobial compounds not only help build the flavor and taste but also protect people from various infectious diseases. Another essential ingredient in the Guilin noodle is monk fruit—a local species (*Siraitia grosvenorii*) of the squash family that bears two- to three-inch round, smooth fruits on its long vines. Monk fruit has long been considered a panacea for treating heat stroke, fever, cough, and sore throat. Local people even believe that it contains an elixir of youth and longevity. Guilin produces more than 100 million monk fruits in its misty mountainous regions each year, which are used mostly as medicines. Non-sugar sweet molecules called mogrosides have been identified in monk fruit, and they were found to be at least 250 times sweeter than table sugar (sucrose), comparable to most of the super-sweet synthetic sweeteners. As a result mogrosides have become a popular sugar substitute and are used to make low-calorie sweeteners.

I have eaten many versions of rice noodles in my life, such as Thai's pad thai and Vietnam's pho. Guilin rice noodles are definitely one of the best. Their high quality, say Guilin people proudly, is because of the superior water from Li River used in making the noodle—a familiar echoing of what the French say about their champagnes and Italians about their balsamic vinegar. Yet the Guilin rice noodles' high quality is also the result of relentless efforts by generations of noodle chefs in Guilin and the continuous demands for better noodles from their customers. Together they let this bowl of noodles invented more than 2,000 years ago endure and still thrive today.

Sushi

A PROPER PLATE of sushi looks more like a piece of art than food; eating sushi is a gastronomic pleasure as well as an aesthetic experience. Sushi is a kind of minimalist food that is stripped to its bare essentials. Even sushi restaurants often look plain and stark so nothing would distract you from the food itself.

Two essential ingredients of sushi are rice and seafood. We tend to think that the seafood is the star of sushi and a piece of fresh tuna or salmon would make a delicious piece of sushi when fastened to a morsel of rice, but sushi masters do not agree. Takashi Saito, the chef of Sushi Saito, a three-Michelin-star restaurant in Tokyo, said: "On the scale from one to ten, the importance of rice is eight to ten." Jiro Ono, a sushi master featured in the film *Jiro Dreams of Sushi*, said: "The most important thing of making sushi is to create a union between the fish and the rice. If they are not in a perfect harmony, sushi will not taste good." In Jiro Ono's sushi restaurant, rice is meticulously cared

SUSHI

for; only the best brand of rice is used and special techniques are developed to cook it. As shown in the movie, a heavy lid that requires two hands to lift is put on the top of the pot in which rice is cooked. On the top of the lid, another pot filled with water is added. Jiro said that his rice needs a lot of pressure to be cooked perfectly. Once the rice is steamed, it is transferred to a wooden tub called a hangiri. While adding rice vinegar slowly to the rice, the cook cools the rice using a hand-held fan and blends vinegar thoroughly into the rice using a wooden paddle called a shamoji, so that the grains are well separated and each one is wrapped with a thin film of the acid. The cooked rice is then kept at ninety-seven degrees Fahrenheit before it is served, because it is believed that rice tastes best at body temperature.

Sushi rice is a short-grain variety that belongs to the Japonica subspecies (*Oryza sativa* subsp. *japonica*). Its grains are short and plump. One of the essential qualities of sushi rice is its stickiness, so that rice grains can be compressed into a lump, but sushi rice is not as sticky as glutinous rice, so each grain in sushi still has its individual identity. I like to watch Takashi Saito making sushi. As shown in a YouTube video, he has a serious look on his face and keeps his eyes laser-focused on the sushi he is making, as if he is under a spell. The swift and fluid movement of his hands and fingers results in a perfect piece of nigiri, and no single grain goes astray.

THE STORY OF RICE

Both seafood and rice, sushi's two essential ingredients, are tied to Japan's unique geography and landscape. Japan is an archipelago comprised of a string of four large islands extending from north to south and thousands of scattered miniature islands. As a result Japan is surrounded by the seas, and it has thousands of miles of shorelines with bays, wetlands, and estuaries—the homes of diverse and abundant marine life. Japan's lakes and rivers are also rich in freshwater fish and shellfish. Its streams and tributaries turn red with salmon when they are returning to their birthplace to spawn. Japanese have been harvesting these natural resources since ancient times, harpooning tuna in the open seas, catching fish in the shallow waters, and gathering shellfish in the intertidal zones.

On the one hand Japanese are lucky people, blessed with the bounty in their waters; but on the other hand, they have also suffered from the destructive power of its waters, as depicted vividly in the iconic image created by Japanese artist Hokusai in his *The Great Wave off Kanagawa*, in which a monstrous wave towers over three fishing boats off the Japanese coast, illustrating the precarious life of the fishermen in Japan.

Japan's land is mostly mountainous. As those rugged mountains are not suitable for living and farming, Japan's villages and towns have grown like vines along the coast. Fourteen percent of Japan's coastal plain is used to grow agricultural crops, and about half of it is devoted to rice. Rice farming in Japan is greatly facilitated by the perpetual streams and springs cascading from its wet mountains to the coastal lowland, where waters are diverted to the rice fields before merging to the seas, so a large portion of Japan's farmland is partitioned into geometric rice paddies, weaved with canals and ditches, and surrounded by dams and dikes. Because rice is the quintessential food for Japanese and rice paddies are ubiquitous in the landscape of Japan, Japanese named several of their cars after the image of a rice paddy—田. Toyota (丰田) literally means a bountiful rice paddy, and Honda (本田), an original rice paddy.

Although Japan is a land of rice today, its rice farming actually started relatively late (around 400 BC) as compared with its neighbors in East Asia. China, across the sea, started to domesticate and cultivate rice at least 7,000 years earlier. Korea, its closest neighbor across Tsushima Strait, has grown rice since 2,200 BC. Why were Japanese so behind in rice farming? One of the reasons lies in the fact that Japan's early inhabitants enjoyed a rich source of foods not only from Japan's super-productive waters but also from its abundant forests—evergreen trees in the southernmost island, deciduous in the middle two islands, and coniferous in the north. As revealed by several archaeological excavations in Japan, the earlier Japanese foraged all kinds of nuts, fruits, and roots from their forests. They also hunted wild animals such as boar, deer, and mountain goats that roamed and lurked in those forests. The ancient Japanese therefore had a diverse and nutritionally balanced diet and saw no advantage to adopting rice farming from their neighbors, especially when rice farming then was relatively unproductive in dry fields, and the rice varieties were poorly adapted to the cool climate in Japan.

Why then did Japanese start rice farming around 400 BC? What led them to change their minds finally? There were a few factors that might have contributed to the shift. For example, farming tools such as hoes and shovels needed for rice farming became more advanced and made farming easier and more efficient. In addition, the development of irrigated rice paddies and cold-tolerance varieties of rice dramatically increased the productivity of rice. As a result rice farming yielded more edible calories than hunting and gathering could, which finally tipped the balance toward farming and triggered the transition to rice agriculture around 400 BC in Japan. It has been hypothesized by some researchers that Japanese adopted the whole package of rice farming—the paddy, the techniques, and the cold-resistance rice varieties—from Korean immigrants at the time.

How did a crop originally domesticated from a tropical grass in southern China start to grow in the cold region such as northeast China, Korea, and Japan? In the history of agriculture, it has always been difficult for a domesticated crop to spread along north-south axes, because the environmental factors such as temperature and day length vary greatly from north to south. Rice was migrating very slowly from southern China to cooler areas as the cold-tolerance varieties were being selected by farmers. Today cold-tolerance varieties are grown in many temperate regions around the world, including California, Spain, Italy, and Portugal. All of them are similar and belong to the Japonica subspecies of rice.

Rice researchers have been investigating the genetic changes underlying the adaptation of rice to cold climates and the northward expansion of rice. They have identified the genes that are associated with cold tolerance. Some of these genes in Japonica rice have diverged from the ones in Indica rice by gaining critical mutations that might have been responsible for its enhanced cold tolerance. Because of these mutations and possibly other unknown mutations, we have the rice for making sushi, risotto, paella, or just plain steamed rice today, all of which are made ideally with Japonica rice.

Today sushi is always connected to Japan, but the ancient form of sushi was not created in Japan. It was originated from a Southeast Asian dish called narezushi, in which salted fish was stored in fermented rice to prevent the fish from spoiling, but it was in Japan that narezushi evolved into a multifaceted gourmet food. Both rice and seafood—the two original ingredients—have remained while others were being invented and gradually introduced to sushi. Pickled ginger was added to cleanse the palette between bites so that you could discern the unique and subtle taste in every piece of sushi. Wasabi, another innovation, was added to boost the flavor and make raw fish safe to eat, with its sharp taste and antimicrobial properties. Additionally two condiments—acidic vinegar and salty soy sauce—have become essential to balance the flavor and taste of sushi.

Sushi is more than just food and its ingredients. It also embodies Japanese culture of perfectionism and work ethic. Generations of sushi chefs in Japan have kept improving recipes and refining techniques relentlessly, bringing sushi from its crude form to a work of art today. Some people may think that sushi is more about technique than creativity and believe that sushi should be kept pure and simple, so there are not a lot of things that a chef can or should do with sushi. The true sushi masters do not agree. Jiro Ono said that he dreamed about sushi every day and jumped out of the bed with fresh ideas. He was still inventing new dishes and improving old ones when he was in his eighties. His devotion is a typical example of the *shokunin*

SUSHI

(craftsman) spirit of Japan, best described by Captain Nathan Ulgrin in the movie *the Last Samurai*: "From the moment that they wake, they dedicate themselves to the perfection of whatever it is that they pursue." In Japan, to learn the skills of sushi making, young sushi apprentices have to endure at least ten years of torturous training before they become a sushi chef, starting from hand-squeezing hot towels for the customers and cleaning fish to massaging octopus and preparing rice. After ten years, they may be allowed to cook egg sushi for customers. In *Jiro Dreams of Sushi*, an apprentice made more than two hundred batches of egg sushi, and all of them were rejected mercilessly. He cried with joy when his egg sushi was finally accepted and when he was praised by Jiro Ono as a true *shokunin*.

I came to like sushi and appreciate its simplicity and subtlety only recently. Sushi is not something you want to eat when you are craving a big flavor or strong taste. Sushi is imbued with subtle and delicate flavors—the mild acidity of the vinegar rice, the umami of the seafood, the slight saltiness of the soy sauce, and a kind of elusive and mysterious flavor that is hard to identify. Sushi lovers covet not only its flavor and taste, but also its texture—the softness of rice, the tenderness of a tuna and salmon, and even the rubberiness of an octopus. Eating sushi is an experience that demands your full attention and mindfulness.

Although I enjoy eating sushi very much, nowadays I eat it only as a rare treat, like going to a concert or taking a vacation, after learning that overfishing has brought bluefin tuna and other marine species close to extinction. I hope we eat sushi sensibly and selectively so that bluefin tuna will not disappear, and our children's children will still have a chance to enjoy its delectable pink muscles designed for deep-ocean swimming.

A Bowl of Gumbo

I EAT RICE almost every day. I like to prepare it the way I had it when I grew up: steamed, served in a bowl, and accompanied by a few dishes—fragrant stir fries or hearty stews. As a Chinese immigrant I blended in well in other areas, but my Chinese stomach still resisted any change. I was entirely oblivious to how other people cooked and ate rice until a dining experience changed my thinking and made me see the tiny grain in a new light.

It all started with a bowl of gumbo.

On an autumn day years ago, I met with several friends for a lunch date at the Olde Pink House, a famous restaurant in Savannah's Historic District, an area bustling with tourists and draped in history. We were seated in a small room with only a few closely spaced tables. While sipping iced tea I looked around at the white-clothed wooden tables, the old portraits on the wall, and the Georgian-style staircase, a setting evocative of the colonial time and style. I scanned the menu carefully as if I were making an important life decision. I ordered a bowl of gumbo—a traditional Southern stew I had heard of but never tried. When the food arrived I saw in front of me a hefty bowl of rice different from any I had eaten until then. It was deep red and thick, and it had an unfamiliar aroma.

Eating my first bowl of gumbo was a revealing experience. It was one of those moments in life that changed me a little, expanding my world slightly. First I looked up recipes for gumbo dishes and found their common ingredients—rice, okra, and palm oil—and cooked a batch at home. I thought about how these ingredients with faraway origins were brought together into the same pot and found their way to the Southern dinner table. One thing led to another, and before I knew it, that bowl of gumbo was leading me on a delightful journey to one of Savannah's older landscapes.

On an early summer day I found myself at the Savannah National Wildlife Refuge—a vast area of wetland in Coastal Georgia and South Carolina. The Savannah River runs its thirty-eight-mile course through the refuge on its journey to the Atlantic Ocean, nourishing the land with its waters.

A BOWL OF GUMBO

Walking on the earthen dikes that enclosed the freshwater impoundments in the refuge, I spotted a few alligators swimming in the stagnant waters and muddy canals. Long-legged waterfowl stood or waded elegantly in the shallow pools. Birds flew high in the open sky, flitted through the grassy marsh, and perched on the leafy trees. The whole area was alive with delightful avian voices. The place was as wild as wild could be. I was scared out of my skin when I saw a two-foot-long snake lying coiled on the dike, motionless, just a few feet from me. I imagined what else could be lurking in the deep swamps and rustling in the dense thickets.

I had come to see the refuge's freshwater impoundments—there are eighteen of them, and they cover about 3,000 acres altogether. These lands were former rice fields during the eighteenth and nineteenth centuries. The elaborate fifty-mile-long system of dikes laced around them was originally built to dam the tidal freshwater pools to build paddies. Wooden trunks were installed along the dikes as floodgates between the paddies and the river. Their simple and ingenious design offered control of the flow of water in and out of the fields.

During high tide, as the salt water moves landward, its higher density allows it to slip under the Savannah River, which raises the water level of the river, its tributaries, and canals in the coastal area. Rice farmers exploited this tidal power to control the

THE STORY OF RICE

level of water in the paddies. They simply raised the trunks during high tide to allow river water to flow into the paddies. To drain, the gates would have been raised during low tide.

The wooden trunk at the Savannah Wild Life Refuge. It was installed along the dike as a floodgate between the paddy and the river.

Standing on the dike alone, with dragonflies darting and mosquitoes buzzing around me, I looked down at the high marsh growing in the impoundment. I imagined the rice field it once could have been and reconstructed every stage of its growth: rows of tender seedlings in spring, waves of hefty stalks in summer, heads of amber grains in autumn, and even the bleakness of bare earth in winter. It was a different

A BOWL OF GUMBO

landscape then, a tilled land created and maintained by the ingenuity and relentlessness of people, at many times involuntarily and with oppressive cruelty. Most of the people working on the rice fields were slaves from West Africa who labored and sweated on this land for centuries, clearing swamps, constructing miles of dikes and canals, and tending the land.

The Savannah National Wildlife Refuge offers only a glimpse of the many former rice fields that dominated the landscape of the Low Country—a land intricately veined with tidal rivers and creeks. Now all of those plantations are long gone and reclaimed again by nature. Nevertheless, that brief history of rice cultivation has left its traces and marks. Old trunks made of ancient wood are still visible, half-buried in the bramble along the dike. Brick cisterns used by the slaves to store drinking water are still intact in the ruins of an old rice plantation. Excavated pottery shards and soda bottles still echo the life of people during rice-plantation days.

One of the legacies left by rice plantations is the trunk, the device that regulated the water level in rice paddies. They are still being used today in the refuge to flood or drain the impoundments periodically, creating diverse habitats for wildlife—moist soil, aquatic pools, marsh, and bottomland.

THE STORY OF RICE

Starting with that day at the Olde Pink House, I came to understand why rice had been woven into the history of the Low Country and how it became an essential ingredient in many Southern dishes—chicken bog, hoppin' John, and jambalaya, to name a few. My inquiry didn't stop there. I later came to know that gumbo's red color comes from palm oil, an ingredient brought by the slaves to the New World. I found out that okra, used to thicken the stew, had been grown under the warm sun in West Africa for centuries before it made its way to the gardens of the American South. Gumbo is indeed a history-laden dish that embodies many narratives.

These seemingly insignificant experiences—a meal, a hike, and a history lesson—helped me connect more deeply with the place I had chosen to call home. I feel more attached to Coastal Georgia after getting to know its unique landscapes, culture, and heritage.

I have come to love this Southern stew called gumbo. Sometimes I cook a pot of it for my family and friends as a special treat.

When Rice Met Soybeans

WHEN OUR ANCESTORS began to select and grow plants for food thousands of years ago, they invariably chose fast-growing and prolific grass species, resulting in the domestication of many cereal crops in different parts of the world, including wheat and barley in the Fertile Crescent, rice and millet in China, corn in Mexico, and sorghum in Africa. These grain crops have become the major sources of food since the dawn of agriculture and account for at least half of the calories consumed by humans today.

Grains generally store starch, which is an ideal source of quick energy. When eaten, starch is rapidly broken down into sugar molecules that fuel all of our biological activities, which is why a long-distance runner often eats a starchy meal before a run or even packs a starch-rich snack for an energy boost during the run. However, in addition to extracting energy from food, we must also acquire a variety of essential substances from the diet to construct and maintain our bodies. Among those substances are amino acids, which are needed as building blocks to make our own proteins that carry out various biological jobs, from transporting oxygen and catalyzing chemical reactions to forming muscles and building bones. Although there are only twenty amino acids, they are combined in various ways to give rise to tens of thousands of proteins, just as the twenty-six letters of the alphabet generate millions of English words.

Among these twenty amino acids, we can make eleven of them from scratch, but have to obtain the other nine, called essential amino acids, from protein-rich foods. Grains, however, generally speaking, are not only low in protein but also lack some of those nine essential amino acids, so if we eat grains mainly but have few high-protein foods in our diet, we will not be able to get enough essential amino acids to make our own proteins.

Our ancestors found a way to balance their diet by domesticating legumes (or pulses) along with cereal crops. The seed of legumes often contains more than 25 percent of proteins. When eaten together, grains and legumes provide a nutritionally complete and balanced diet. Various grain-legume combinations have been adopted around the world, including wheat and barley with peas and lentils in the Fertile Crescent, corn with beans in Mesoamerica, and rice with soybeans in China. The incorporation of legumes into the diet has not only ensured our survival, but also shaped our traditional foods and regional cuisines. Today hundreds of lentil dishes in the Mediterranean region and West Asia or refried beans in Mexico or soy sauce and tofu dishes in China are examples of the culinary legacies left by the adoption of legumes in the past.

When rice and soybean plants are placed next to each other, they look like an odd pair. Botanically speaking rice and soybeans are different in every possible way, because they represent the two major branches of flowering plants—monocots and dicots, which split from their common ancestor about 140-150 million years ago and diverged considerably since then. Soybeans are a dicot plant, meaning that the seed has two cotyledons—the two familiar halves in a bean or pea, whereas rice is a monocot and has only one cotyledon in the seed. Additionally, soybeans have broad leaves and netted veins, in sharp contrast to rice plants with narrow leaves and parallel veins. In soybeans two to four oval-shaped seeds are snuggled within a pod when mature, and all the major elements of a soybean seed are orderly arranged and are identifiable even with a naked eye—its thin coat, its two cotyledons, and its baby nestled between them. But the anatomy of a rice seed is little muddier; every part of it appears in a small format and the boundary between the parts is blurred. The cotyledon in a rice kernel is miniscule and invisible. Instead, the bulk of a rice kernel is a structure called endosperm

THE STORY OF RICE

that stores food for its embryo. Yet an important distinction that interests us is that soybeans and rice pack different types of foods for their babies in the seed—a soybean plant fills the two cotyledons with proteins (about 38%), oil (about 20%), and other micronutrients, whereas rice loads its grains primarily with starch, so a diet made of rice and soybeans is both energy rich and nourishing.

Since the domestication of soybeans, they have been a major ingredient of the diet in Asia. They are often eaten as edamame in Japan or China, which is beans-in-the-pod gathered before the seeds are fully ripened. Edamame is green, tender, and sweet. It also has an umami taste because of the presence of flavorful amino acids in the young beans. Edamame is often served as a side dish or snack with a sprinkle of salt or a pinch of sesame seeds after being boiled briefly. As beans are shelled at the very last second, literally, edamame is fresh and fragrant—a delightful treat. In addition to edamame, soybeans are often made into soymilk by simply soaking seeds in water; blending them into a smooth, foamy liquid; straining the liquid; and finally boiling the collected liquid. Today soymilk can be made in a machine with the press of a button. Homemade soymilk has a natural "beany" flavor that is often removed from soymilk sold in stores under names such as Silk and Vitasoy.

In East Asia or Southeast Asia, however, most of the soybeans are made into tofu, which is made by coagulating soymilk into curds that are then pressed into solid white blocks with various textures, such as silken, soft, firm, extra-firm, or dry, depending on how much water is pressed out. Based on archaeological data, tofu was invented during the Chinese Han dynasty about 2,000 years ago, and then it spread to other parts of East and Southeast Asia. According to a popular Chinese legend, Prince Liu An (179 – 122 BC), who lived during the Han dynasty, accidentally added salt to his soymilk and caused it to coagulate into curds—a happy accident that resulted in one of greatest culinary inventions in history. Today several agents, including salt, acid, or an enzyme, can be used as a coagulant that transforms liquid soymilk to solid curds in the blink of an eye. Chemically speaking a coagulant destroys the natural structure or shape of protein molecules in soybeans, uncoiling or untangling them, causing them to lose their solubility in water and aggregate into solid form—a process similar to the boiling of an egg that turns it from fluid to solid.

Freshly made tofu has a subtle flavor and accommodating nature. It can be served as a cold dish, either sweet or salty, with just a little seasoning. It can be stir fried, deep fried, braised, baked, steamed, grilled, or stewed. Because of its porous texture, it absorbs

THE STORY OF RICE

flavors well, so it is often seasoned or marinated to a desired taste and flavor. Hundreds of dishes are built around tofu, and new ones are still being invented. Some examples of homey and mouth-watering tofu dishes are crispy tofu fried in hot oil, soft tofu stewed with meat and vegetables, crunchy tofu roasted on a charcoal fire, or rubbery dry tofu stir fried with vegetables. Because tofu is spoil-prone at room temperature, people often freeze tofu to preserve it. After it is thawed, tofu can be cut into cubes and stewed in a pot of soup with mushrooms, meats, and other ingredients of your choices. Because tofu becomes very porous and spongy after the freeze-thaw cycle, it holds more broth, so when you bite it, the fragrant soup gushes out and spreads across your mouth—an incredible sensation.

Mapo tofu, however, is the most well-known Sichuan-style tofu dish. The name of this dish sounds comical in Chinese: "ma" stands for pockmarks, and "po" means an elder woman or grandma; hence mapo tofu is also translated to "pockmarked grandma's tofu." In this dish tofu is cooked with minced pork meat, flavored with a concoction of spices and condiments—rice wine, garlic, ginger, Sichuan peppercorn, and fermented bean and chili paste, just to name a few. The finished product is a heap of cube-shaped tofu suspended in a thick layer of bright red soup and garnished with a sprinkle of green scallion. An authentic dish of mapo tofu is supposed to be numbing, spicy, tender, and aromatic. However, a true classic mapo tofu dish is hard to find outside of Sichuan nowadays. Many restaurants serve mapo tofu of their own versions and interpretations that vary widely from region to region, but Sichuan peppercorn and chili bean paste are the two fundamental ingredients of mapo tofu—Sichuan peppercorn for its tingly and mouth-numbing effect and chili for its fiery spiciness. There is no mapo tofu without them.

No matter how mapo tofu is cooked, it must be served with a bowl of steamed rice. Eating mapo tofu without rice is a blasphemy against the sacred culinary tradition. Chen Kenichi, the famous Iron Chef, nicknamed Sichuan Sage on the television series of *Iron Chef*, cooked mapo tofu twice in slightly different styles in episodes that featured tofu as the theme ingredient. Both times, he served mapo tofu with a bowl of steamed rice, and both times, he won. Anthony Bourdain, an American celebrity chef, praised mapo tofu as the apex of Sichuan cuisine in the episode "Sichuan with Eric Ripert" in *Parts of Unknown*—an American travel and food show. Yes, both he and Eric ate mapo tofu with a bowl of steamed rice—there is just no any other way to eat mapo tofu. The pairing of mapo tofu with rice is considered one of the best illustrations of the yin-and-yang philosophy in Chinese cuisine. Mapo tofu and rice are in contrast and yet complementary in every way—taste, color, flavor, nutritional profile, and cooking method. The combination results in a balanced meal, with a complex and explosive taste of mapo tofu accompanied by the simple and subtle taste of rice.

In addition to freshly made snowy white tofu, soybeans are often fermented to produce flavorful condiments or food. Soy sauce, a brownish liquid condiment, is made mainly from fermented soybean paste. Invented about 2,200 years ago during the West Han dynasty of ancient China, soy sauce spread to other parts of Asia. Since then soy sauce has become the foundation of the East Asian and Southeast Asian cuisines. A dish flavored with soy sauce is almost automatically identified as Asian. Miso, a traditional Japanese seasoning, is produced by fermenting soybeans and a few other ingredients. Tempeh, a cake-like food in Southeast Asia, is fermented tofu that was invented in Indonesia. Just like many other fermented foods around the globe, some of the fermented soy products are challenging to the senses. For example, natto, a famous Japanese delicacy, is a fermented soybean dish with a disgusting slimy texture and repulsive ammonia smell. Stinky tofu, a Chinese tofu dish, reeks of rotten egg and decaying meat. Hairy tofu in China is covered with a layer of white fluff reminiscent of molds on old bread or oranges—a sign of rot and decomposition. However, these visually and olfactorily repulsive soy products often taste delicious, because the microbes in the environment have turned proteins and fat into amino acids, esters, or other flavorful molecules during the process of fermentation. It is intriguing how something that smells so bad can taste so good—a sensory paradox that is mystifying and fascinating. As with many other infamous stinky foods, such as Limburger cheeses in Europe or lutefisk in Norway, these delicious-disgusting soy foods appeal to many people. Once you acquire the taste and embrace the paradox, you are truly in heaven.

All in all, soybean-based foods, whether fermented or not, have enriched the diet and delighted the palate of many generations of people in Asia. Soybeans have accompanied rice for centuries or even millennia, whether as a small tray of fermented tofu with rice porridge at breakfast or hearty mapo tofu on a heap of steamed rice at dinner. Rice and soybeans are central to the cultural identity for many Asians. Recently I watched *In This Corner of the World*—a 2016 animated Japanese drama. During World War II, rice was strictly rationed and tofu or other soy foods were scarce. In the story, Suzu, an innocent and kind-hearted young wife, managed to put food on the table using a cooking technique to make rice grains swell and appear larger. Her mother-in-law stashed away a small bag of rice and saved it as the last meal for the family when the time came. In a desperate moment when the last hope slipped away and all efforts seemed to end in vain, Suzu tried to gather her last strength by telling herself that her body was made of rice and soybeans and reminding herself who she was.

THE STORY OF RICE

That statement resonates with many Asians who have been sustained and nourished by rice and soybeans.

The Western world never heard about tofu until the late eighteenth century. Benjamin Franklin was the first American who mentioned tofu. In his letter to James Flint, a British merchant who was responsible for the introduction of soybeans to North America, he discussed how soybeans were turned to tofu. In another letter to John Bartram, an American botanist, he referred tofu as cheese from China. Benjamin Franklin was absolutely right, as the science and technique behind the cheese- and tofu-making are essentially the same. Both use a chemical agent to coagulate proteins; both are often preserved and flavored through fermentation; and both have unique local taste and flavor because of a distinctive combination of the regional climate and microbial community. Italo Calvino said in his book *Mr. Palomar*, "Behind every cheese there is a pasture of a different green under a different sky." The same statement holds true for tofu, and every tofu is a unique product of geography, culture, and human ingenuity.

Soy-based foods have recently become popular among vegans and vegetarians who regard them as healthy alternatives to meat and dairy products because of their high protein content and presence of all nine essential amino acids. There are some debates about the possible role of soy products in the development of cancer, however. Some researchers believe that soybeans contain isoflavones that are chemically similar to estrogen—a female sex hormone. As estrogen is linked to breast cancer in some cases, researchers argue that soy-based foods may increase the risk of developing breast cancer. Other researchers reason that Asian women have been consuming soy foods since ancient times, yet the rate of breast cancer has been generally low in Asian countries. According to the American Cancer Society, "The evidence does not point to any dangers from eating soy, and the health benefits appear to outweigh any potential risk. In fact, there is growing evidence that eating traditional soy foods such as tofu, tempeh, edamame, miso, and soymilk may lower the risk of breast cancer, especially among Asian women." Its statement is backed by many studies, although scientists still don't know the biological mechanisms by which the consumption of soy foods reduces the rate of breast cancer.

Shaoxing Rice Wine

ALMOST EVERY CULTURE in the world invented its own version of wine, beer, or both using yeast and local fruits, grains, or tubers. This fact is exemplified by sorghum beer and palm wine in Africa, potato chicha (a type of beer) and pepper berry wine in South America, barley beer and grape wine in the Middle East, and rice wine in China. Analyses of the chemical residues from a clay jar unearthed at the Jiahu archaeological site in China confirmed that ancient Chinese were making an alcoholic drink by fermenting a mix of rice, grapes, hawthorn berries, and honey, literally a cocktail of beer, wine, and mead, about 9,000 years ago—one of the oldest known deliberately fermented beverages.

Rice wine had been the drink for Chinese for thousands of years before grape wine and beer were introduced to China. It has been associated with many traditions, folklores, and legends. Drinking rice wine is a familiar scene or ritual often featured in Chinese classic literature and poetry. For example, *The Water Margin* (or *Outlaws of Marsh*), one of the four great Chinese classic novels, depicts many memorable and iconic drinking scenes. It is a story about 108 outlaws in the Song dynasty (960 – 1279) who are granted amnesty by the government and enlisted to fight foreign invaders and internal rebels. In the novel, as rice wine flows to the characters' blood streams, they come alive. A friendship is forged, a dispute settled, or a crime committed over a few bowls of rice wine. Wu Song, one of the major characters in the book based on a real person, is turned into a daredevil by alcohol. After gulping down eighteen bowls of rice wine in a village tavern, he crosses a ridge, against all the warnings, to a dangerous area where tigers are lurking. When he encounters one, instead of running for his life, Wu Song knocks the tiger down, pins it to the ground with one hand, bashes its head repeatedly with another hand, and slays the tiger in his drunken rage. In the era of no wildlife protection laws or acts, tigers were viewed as monsters that attacked humans and livestock. Wu Song thus has been hailed as a hero, and his tall tale has been told for many centuries. In December 2020, Netflix announced that it would be adapting *The Water Margin* into a movie. The news caused both enthusiasm and concerns among many Chinese. Some people thought that the ancient values and cultures in China are so alien to Western audiences that they might be lost completely in translation, whereas many others were excited with the anticipation of seeing an old story being told in a new light.

Like the rest of the world, wine and poetry are inseparable in Chinese history. Just as the imagination of Western poets has been ignited by fermented grapes, the mind of ancient Chinese poets had been stimulated by fermented rice. Many beautiful poems were written by poets who were either rapt with wine or in a drunkard state. Among them, Li Bai (701 – 762) of the Tang dynasty stood out a great poet who wrote more than one thousand poems in his life. The famous bard indulged his love of wine and was also captivated with the dream-like image of the moon. Both wine and moon found their way into many of his best verses. He is often depicted in paintings as a solitary man raising his bowl to a bright full moon in a clear sky. In one of his most famous poems, "Drinking Alone beneath the Moon," he personalized the moon playfully as one of his drinking companions, exuding his loneliness and sadness in the opening lines of the poem.

> *A jar of wine, in the midst of blooms,*
> *I drink alone, with no dear friends.*
> *Raising my cup, I invite the bright moon,*
> *The moon, me, and my shadow are three drinking companions.*

Wine opened the floodgate of Li Bai's imagination and evoked in him a strong sentiment. The line

SHAOXING RICE WINE

between reality and fantasy was blurred in his poetry and in his life. According to a legend, while drinking in a boat and intoxicated, Li Bai tried to embrace the reflection of the moon in the placid lake, fell into the water, and drowned.

Just as many famous wines are often associated with particular places, rice wine is tied to Shaoxing, a city of the Zhejiang province in China. Shaoxing is a water town nicknamed the Venice of the East. Its landscape is crisscrossed by rivers, waterways, and canals and strewn with lakes and ponds. In old times people often got around by a wupeng—a boat with a black awning, equivalent to a gondola of Venice in its functionality. Shaoxing's dry lands are connected with numerous bridges—at least hundreds of them. Those bridges have not only provided the safe passage across rivers and connected the community, but also revealed something about the identity and ingenuity of generations of Shaoxing people. The city is literally an unroofed museum of bridges—bridges of all sizes, shapes, materials, and styles. Some of the stone bridges are thousands of years old, still standing as silent witnesses of Shaoxing's history. A part of Shaoxing's purest water has been turned to Shaoxing wine—a yellow-tinted rice wine—during the past thousands of years. Shaoxing wine is the most sought-after rice wine in China and other parts of Asia, and some of its famous brands have beautiful and poetic names such as Snow-flavored, Rosy Daughter, and Engraved Flower.

Besides its bridges and rice wine, Shaoxing is also known as one of oldest cities in China, with 6,500 years of history. It boasts of having more scholars or other notable people in Chinese history than any other city in China. One of them is Lu Xun (1881 – 1936), the greatest writer and the founder of modern Chinese literature. Many of Lu Xun's stories were set in a town called Lu Zheng, which was really a fictionalized version of Shaoxing, his hometown. Several of the most memorable scenes in his stories happened in Xianheng, a rice wine tavern owned by one of his uncles. He liked to use the tavern as a window to peek into society and the lives of ordinary Chinese people of his time.

A character featured in one of his short stories was implanted into my mind and perhaps everyone's mind of my generation. It is "Kong Yiji," which found its way into the textbook for high school literature class during Mao's era. The story is set in the Xianheng tavern where wine is served in an old-fashioned way—ladled out of a big jar into a bowl and warmed by hot water and often served with a dish of salted bamboo shoots or flavored aniseed beans. Kong Yiji is a scholar who has repeatedly failed the imperial examination—a qualifying test for civil service jobs in the Qing dynasty.

SHAOXING RICE WINE

As a result he can't secure a single job. He is also an unskilled book thief and always gets caught in the act. But whenever he has a few coppers (Chinese coins) to spare, he comes to the tavern for a drink. In the tavern he becomes a figure of fun and is often mocked by the fellow customers. In the end he is caught stealing again and is beaten until his leg breaks. He drags himself all the way to the tavern to have a drink and is ridiculed mercilessly again. Although as a schoolgirl I didn't fully understand the deep social meaning conveyed by Lu Xun, I still remember, forty years later, the sadness I felt when I read "Kong Yiji drags himself out of the tavern and is never seen again." Lu Xun created a timeless figure—a victim of corrupted society and human cruelty who eases his pain and seeks comfort with alcohol. Drinking, however, only worsens his misery and leads him to a more perilous path and deeper trap.

In 1981, years after its closing, Xianheng was reopened in a grand style to the public to commensurate the one-hundredth anniversary of Lu Xun's birth. In the front of the building a statue of Kong Yiji wearing a traditional gown and a long braid characteristic of the Qing dynasty was erected. A bowl and a plate placed next to him are symbolic objects representing the wine tradition of Shaoxing: a bowl of rice wine paired with flavored aniseeds.

THE STORY OF RICE

To make alcohol from grains or tubers, starch must be broken into sugars that in turn are fermented into alcohol by yeast. Various cultures around the world have invented their own ways by which starch is converted into fermentable sugars. In the West, grains are first sprouted (or malted), and starch is broken to simple sugars by natural enzymes in the sprouting seeds. In some parts of the world, grains or tubers are chewed in mouths. Although it sounds unappetizing, amylase—the enzyme in saliva—accomplishes the same task, turning starch to sugars. In Asia special molds, such as *Aspergillus oryzae,* were discovered to digest starch in rice grains. Jiuqu (or simply qu), the fermentation starter used for making rice wine, is a dry mixture of molds and yeasts, all in suspended animation. During fermentation they come alive when mixed with water and exposed to warm temperature. The enzymes in the molds break down the starch first, and then the yeasts take over, fermenting the sugar into alcohol.

Technically rice wine actually is beer, as it is fermented from grains instead of fruits. Making rice wine is a long and tedious process. Traditionally people grow rice, mostly a variety of sticky rice, in the spring and make wine in the winter when water is at its purest. Grains are washed, soaked in water, dried, and steamed. The cooked rice is then transferred to a temperature-controlled room. As the rice cools, qu is scattered and kneaded into it. Pure water is then

SHAOXING RICE WINE

added to create a mixture that is then fermented for twenty-five to thirty days in a big vat. Brewers tend to it around the clock, checking its smell, color, consistency, and many other subtle qualities. When the yeast is killed by alcohol—its own excretion—fermentation is complete. Liquid is extracted from the mixture through a pressing or straining process and then bottled or stored.

Making rice wine by hand is more an intuitive art than an exact science. There are hundreds of variables in the process—the starter, the temperature, the incubation time, etc. Through years of trial and error at work, a master brewer develops a deep intuition and experiential wisdom. By looking, smelling, tasting, listening, and touching, he can discern subtle things and tiny details in the fermentation vat, all of which inform him about the quality of wine. Although nowadays rice wine is being manufactured on a large scale with automated machinery, it is still made by hand as a time-honored tradition. In Shaoxing and many other cities in Southern China, rice wine is often homemade and used to celebrate births and marriages, commemorate deaths, and forge and strengthen bonds in the family. An old and beautiful tradition has been practiced for many years in Shaoxing: when a daughter is born to a family, the happy father makes three jars of rice wine. He then seals and buries them deeply under an osmanthus tree. Years later, when the daughter gets married, the wine is taken out and served at the wedding banquet. The first three bowls are always presented to his daughter's husband and in-laws, a wish for a happy marriage and harmonious family relationships.

Even though I don't drink full-strength rice wine, I enjoy a scaled-down version of it called rice wine soup. To make it, rice is not fermented all the way, and as a result the product is a mix of sugar with low-level alcohol (1-2%) and lactic acid (0.5%)—a fragrant liquid with fluffy, partially digested rice grains floating in it. The starter for making rice wine soup is often sold in the form of small balls about one inch in diameter in Asian groceries or even online. Just like making full-strength wine, rice is first soaked and steamed. When the rice cools down, a hard ball of starter is cracked into pieces with a knife and then crushed with a rolling pin into a fine powder, which is sprinkled over the rice in a container and mixed with rice thoroughly. Water is then added and blended in. The mixture is left to sit in a quiet place for several days. Whenever making a batch of this dessert, I check and sniff the content often, anticipating the magical transformation. Once I detect a pleasant smell resulting from a balanced mix of sugars, alcohols, lactic acids, and other mysterious aromatic molecules, I halt the fermentation by placing the container in the refrigerator, where it can be stored for weeks or months. I like to cook it with egg drop or a poached egg or other odds or bits like osmanthus flowers and wolfberries, serving it either pepper-hot in the winter or icy cold in the summer. Either way, it hits the spot.

Shaoxing wine is also commonly used in cooking. A bottle of Shaoxing wine is a fixture, along with soy sauce, in every Chinese kitchen. It not only offsets the unpleasant odor in meat and fish, but also adds flavors to an otherwise ordinary food. A well-known Shaoxing dish called drunken chicken is made by marinating freshly steamed chicken in Shaoxing wine overnight in a refrigerator. The next day, the chicken would be permeated with the tantalizing aroma of Shaoxing wine—a delicious cold dish coveted by many Chinese. In addition to drunken chicken, Shaoxing is also known for its other drunken dishes—drunken shrimps, drunken fish, and drunken crabs, all of which are marinated and cooked in Shaoxing wine. Another culinary tradition in Shaoxing is that the dregs—the byproduct of fermentation—are saved to cook many dishes, turning waste into a delicious ingredient in Shaoxing cuisine.

The Science of Making Rice

THERE ARE COUNTLESS ways of cooking rice. The simplest form is plain steamed rice that I grew up with. I love its soft texture and subtle flavor that goes well with any Chinese dish. Steamed rice is made from just two ingredients: rice and water. The recommended ratio of rice to water is close to one to one, which is a good mathematical ratio for a beginner—easy to remember, easy to quantify, and easy to scale up and down for any quantity. Yet sometimes you may have to adjust the ratio a little for a different brand of rice or for a desirable texture. During my childhood my family of three generations ate meals together. My grandparents preferred softer rice because of their chipped or lost teeth, while the kids liked the firmer texture better. To accommodate everyone in the family, my mom shaped rice into a hill with a slight slope from one side to another in the pot. As a result the

THE SCIENCE OF MAKING RICE

rice at the bottom of the hill (with lower rice-to-water ratio) was cooked softer than the rice toward the top of hill. With a little creation and thoughtfulness, my mom made everyone in the family happy.

Once rice and water are mixed in a pot, physics and chemistry come into play. The pot is brought to a boil first and then to a slow simmer. While simmering, water molecules enter the rice grains through the physics law of diffusion, and the grains puff up. In the meantime the water molecules undergo a phase transition from their liquid to gaseous (or steam) form, so the heat is transferred to the grains through the hot steam by the second law of thermodynamics, that is, heat always flows from an area of high temperature to low temperature—the same law that underlies the steam engine or explains the expansion of the universe. When all the liquid is evaporated from the pot, the heat is turned off and the rice is left to rest. It is essential that a lid be kept on the pot all the time to prevent the steam from escaping. In that way, heat is transferred through hot steam, not liquid water, which is why we call it steamed rice. In reality, though, this type of cooking is a combination of steaming and boiling.

In the pre-rice-cooker era, cooking rice needed full attention, especially when cooking on an old-fashioned stove with a live fire, which was exactly how rice was cooked when I was a girl growing up in China. On those old stoves rice could be

THE STORY OF RICE

easily overcooked, undercooked, or even burned. If you happened to lift the lid before the rice was fully cooked, the whole pot of rice was ruined, as it yielded half-cooked rice that couldn't be remedied. I still remember the consequences of an ill-cooked pot of rice—a family bickering or even a domestic drama. It was not just a kitchen accident but also an economic loss when your family hardly made ends meet and rice was strictly rationed.

The invention of modern electric rice cookers by Japanese has relieved many families, especially housewives, from the drudgery and daily time-consuming rice-cooking chore and freed them to pursue other activities. Nowadays cooking rice is just a matter of pressing button, and we rarely give a thought to what is going on inside a rice cooker. In fact all the steps of rice cooking—soaking, boiling, simmering, steaming, and resting—are executed precisely and seamlessly in a rice cooker, guided by an electric heating plate and a thermal-sensing device, resulting in a perfectly cooked pot of steamed rice.

The chemistry of cooking rice is relatively simple because of the chemical purity of rice, which is 85-90 percent starch and only a little protein and other substances. As a result, there is little Maillard reaction (a chemical reaction between proteins and carbohydrates) that gives some foods, such as steak and bread, their desirable flavor and brownish color. Caramelization, another chemical reaction involving

THE SCIENCE OF MAKING RICE

the breakdown of sugars at or above 160°C (or 320°F) and formation of brownish caramel also should not occur in a rice cooker, as it rarely exceeds the boiling temperature (100 °C or 212 °F). However, if rice is cooked on a stove, the excessive heat may cause rice at the bottom of pot to caramelize and result in a layer of hard crust with a nutty flavor and brownish color. Although it is not a desired outcome for steamed rice, the hard crust of rice has found its way into several well-known soups or dishes in China, which were probably created as a result of culinary accidents and the habit of frugality.

What happens chemically in the rice cooker is a reaction called gelatinization. Before being cooked, each grain of rice is packed with numerous granules, which look like a bunch of blobs under a microscope. Each granule is filled with two types of starch molecules: linear amylose and branched amylopectin. Together they form an orderly and crystal-like structure within the grain. During cooking, the water molecules penetrate the granule, seep into the space, and force starch molecules to separate. As a result the granule increases its volume and finally burst opens. When a granule collapses, the linear amylose molecules leak out of the granule, whereas branched amylopectin molecules are bonded with water molecules, forming a gelatin-like mixture. The process is therefore called gelatinization, which gives steamed rice, especially the high-amylopectin variety, a soft or silky texture. You can actually see the swelling and bursting of those tiny granules and releasing of the starch from them if you view it under a microscope. A similar process also operates when we use starch as a thickening agent for making soups or other dishes.

In addition to pure steamed rice, rice is often cooked together with other ingredients. In India rice is mostly boiled in a large amount of water and then drained. As long-grain rice is used, the finished product is fluffy with well-separated grains, which are easily seasoned with herbs and spices to create a flavorful and tasty rice dish.

In Italy risotto is a rice dish cooked with many ingredients using a more complicated technique. First chopped onions, mushrooms, or other ingredients are sautéed in oil or butter in a pot and then raw rice is added to the pot and toasted until a complex flavor is achieved from the Maillard reaction and caramelization of proteins and carbohydrates in the ingredients. A small batch of stock prepared from meat, vegetables, or seafood is then added and rice is simmered in the stock and stirred constantly until water is nearly evaporated. Another small batch of stock is then added and simmered and stirred. The process repeats several times until rice is cooked to its desired outcome—creamy, flavorful, and al dente (firm when

THE SCIENCE OF MAKING RICE

bitten). As many chemical reactions take place during the long period of simmering, a risotto dish has a depth of flavor that comes from the integration of all the ingredients in the pot. Italians believe that the key to a good pot of risotto is the stock; "Without it, there is no risotto." I laughed at a risotto scene in the movie *Big Night*. The Italian chef in the movie is frustrated with the ignorance of an American customer who asks for scallops or shrimp in her seafood risotto and does not know that seafood risotto means rice cooked in seafood broth. The customer also asks for pasta, another starchy food, to go with her risotto—an unforgivable culinary blasphemy to the Italian chef.

Risotto is often cooked with short-grain rice, such as Arborio. Some rice varieties have been specifically developed to make risotto. For example, Carnaroli—one of the varieties developed by Emiliano Carnaroli—is considered the best rice for risotto and is often referred to as the king of Italian rice. As Carnaroli is a medium-grain rice and little less sticky than short-grain varieties, it keeps its shape better once cooked—a perfect balance between absorbing seasonings and retaining consistency during slow cooking. As a result risotto made from Carnaroli has a firm bite and creaminess that Italians covet.

Paella, the national dish of Spain, is similar to risotto in its spirit and science. It starts with making stock from a variety of ingredients of meat, vegetables, and seafood and then simmers rice (ideally short-grain varieties such as Bomba or Calasparra) in the stock until the rice is cooked to its desired outcome. A good pan of paella comes with a layer of toasted rice, called socarrat, at the bottom, which is considered a delicacy and absolutely essential. To get a perfect layer of socarrat, the cook must listen carefully when paella is nearing its completion. Once the aroma of toasted rice wafts to the nose, the pan must be removed immediately from the heat to prevent burning the rice. As many chemical transformations, including the Maillard reaction and caramelization, occur within the bottom layer of rice, socarrat has an incredible depth of flavor and taste.

Around the globe there are many other ways of cooking rice, but no matter how rice is cooked, the right type of rice should be used to achieve the ideal flavor and texture. It would be helpful if the cook knows a bit of the science behind rice and rice cooking, so he or she can bring out the best in the rice and transform it into a delicious dish.

A Leaf of Grass

A LEAF OF grass looks like a blade, long and narrow, and it is streaked with parallel veins from tip to base. Although it looks simple to the naked eye, the leaf is a marvelous biological machine designed to perform an amazing job: photosynthesis.

A grass leaf is as thin as a razor blade, yet it is composed of multiple intricate layers as seen under a microscope. In its upper epidermis, or skin, that faces the sun are a few special cells called bulliform cells, which are clustered together and appear bubble-shaped and empty under a microscope. Those cells help regulate the rolling or unrolling of the blade by gaining or losing water in response to the level of moisture around them. When the water supply is sufficient to the plant, the bulliform cells absorb water and swell, becoming stiff and turgid. As a result the leaf straightens and unfolds, as if stretching itself toward the sun. On the other hand, at a time of low water supply, these cells lose water and become flaccid. As a result the leaf curls inward, less exposed to the dry air and sunlight, a strategy to conserve water during a period of drought. This process of furling and unfurling is governed by the basic principle of osmosis; that is, water molecules enter or leave a cell in response to different concentrations of water between inside and outside of the cell, the same principle underlying the curling of the leaves in a touch-me-not plant in response to a touch or closing the trap in a Venus flytrap when triggered by an invading fly.

Tiny pores called stomata are embedded in the lower epidermis, the lower surface of a leaf, each surrounded by a pair of sausage-shaped cells called guard cells that regulate the opening and closing of the stoma (singular form of stomata) by the same principle of osmosis. Carbon dioxide in the air needed for photosynthesis enters the leaf through those openings. At the same time, precious water also escapes the leaf through them, so a leaf gains carbon dioxide and loses water simultaneously through stomata. Opening or closing stomata, therefore, is a delicate balance that a leaf must maintain by weighing the pros and cons and assessing the loss and gain, especially when the weather is hot and dry.

The most magical cells in the blade, however, are mesophyll cells sandwiched between the two layers of skin. They are solar cells equipped to harness sunlight. A mesophyll cell is filled with thirty to forty green blobs called chloroplasts, each of which contains green chlorophylls that absorb sunlight and transform solar energy to chemical energy.

Sunlight is a mixture of colors with differing wavelengths—an optical phenomenon discovered by Isaac Newton more than three centuries ago when he shone a beam of sunlight on a glass prism and saw a rainbow of hues emerging magically on the other side. Not all the colors of sunlight are useful for photosynthesis, though. A leaf's absorbed colors of light can be deduced from its mirror image—the reflected ones. The green, which often signifies growth and life, actually is the least useful color for photosynthesis, as it is not absorbed, but thrown back to you by the leaf. Other colors, however, such as red or blue, are gathered by the leaf for photosynthesis.

When the sun is up, its light energy is harnessed and used by the chloroplasts to make sugar molecules. Those sugars are then transported to other parts of the plant through those parallel veins, where they could be burned to fuel the growth, rewired to strengthen the defense, or just stored away. In a tree the product of photosynthesis is manifested in the visible world as thick and tough bark, massive canopies, enormous root systems, or bountiful fruits. In a grass such as rice and other cereal crops, most of the sugars made by photosynthesis are stored in their tiny grains—lunchboxes packed for their babies.

THE STORY OF RICE

The leaf is a quantum world, just like everything else in the universe. After a beam of sunlight with its wave-particle duality zips through vast space, its thousands of trillions of light particles called photons hit the surface of a leaf. The energy of each photon then travels through a maze of chlorophyll molecules and reaches its destination, the reaction center in the chloroplast. It was discovered recently that the energy in a photon does not hop randomly through the maze as we used to think. Instead the energy in each photon behaves like Schrodinger's cat and can exist in multiple states and travel in multiple paths simultaneously (known as quantum walk), before it reaches the "reaction center" where the light energy is converted to the chemical energy. Quantum walk allows energy to find the shortest route and arrive at its destination in time. The phenomenon of quantum walk has inspired scientists and engineers to develop solar panels that could do the same trick. Quantum physics is something more than I can grasp, but I am astonished by the intricacy and complexity of a seemingly simple leaf. Indeed, "a leaf of grass is no less than the journey-work of the stars," as Whitman said poetically and metaphorically.

Photosynthesis, like all other biological processes, is consistent with the laws of nature, obeying the rules of chemistry and physics. After all, at a deeper level a leaf is made of the same building blocks—atoms, electrons, and molecules—as everything else in the universe. In fact many physicists and chemists have played seminal roles in elucidating the basic mechanisms underlying photosynthesis, such as the flow of energy, the collision of chemicals, and the excitement of electrons. Photosynthesis is, in its essence, a process of funneling and transforming energy, which can be explained in theory by the law of physics.

Still, a leaf is more than a solar panel or an energy machine. The difference between a living and nonliving thing is profound and mysterious. Unlike a cold and unanimated object, a leaf or any part of an organism, is messy, as it is affected by many variables and unknown factors. Any physicist or chemist who attempts to apply reductionist approaches to explain it would be likely to make an unintended error, sometimes with a disastrous consequence. One such mistake was made by an eminent physicist and rocket scientist in China in the late 1950s and early 1960s.

His name is Qian Xueshen; none of the other physicists in China has ever held a prestigious position or as renowned as Qian. Qian, educated at MIT and Caltech, was a leading scientist in America's missile programs and the Manhattan Project. However, during the Second Red Scare in the 1950s, Qian was accused of being a communist sympathizer and was detained and later deported. Since his return to China in 1955, Qian had been the key player in China's missile and space program and hailed as one of the founding fathers of Two Bombs and One Satellite projects in

THE STORY OF RICE

China. Qian was such a notable space scientist that his name has appeared in several popular science fiction books, including Arthur Clarke's *2010: Odyssey Two*. Without a doubt Qian's scientific contribution was significant, but somehow he stepped into the messy worlds of biology and politics, or rather, was dragged into them during a manic period called Great Leap Forward (1958 – 1961), a movement launched by Mao Zedong and his followers with the grand intention of catching up or even surpassing America's and Britain's economies.

As a scientist Qian was asked to use his knowledge in math and physics to calculate the potential yield of rice and other crops. In several articles he predicted that rice could potentially produce more than 40,000 kilograms of grain per mu (0.16 acre) by maximizing the absorption and conversion of sunlight, about twenty times the actual yield (2,000 kilograms per mu) at the time. Unfortunately his pure theoretical projection was used by Mao Zedong and the party to motivate peasants and set an unrealistic yield target. Furthermore, many local officials, under the pressure from the central government, competed with each other to report their much-exaggerated grain harvests to earn favors. As a result the peasants had to turn in more grain to supply people in towns and cities or even to export, which left almost nothing for the peasants to feed their own families. Starvation set in. Rice disappeared first from the dinner table and then potatoes and soybeans and then vegetables and fruits. Desperate and famished, the peasants dug up roots and bulbs, collected acorns and tree bark, and foraged for wild greens and herbs. The famine raged on until 1962. Tens of millions died of hunger during the three-year famine. Many babies born during this period were undernourished either in the womb, during their early childhood, or both. I was one of them.

The famine was deeply imprinted into the collective memory of several generations of Chinese people. Although Qian's story was only a small interlude in Mao's grand scheme and disastrous agricultural policy that led to the Great Famine, he had been blamed for his role. Many people thought that Qian owed Chinese an apology, but he never did. He died at age ninety-seven in 2009, and the world would never know what he thought of that episode in his celebrated scientific career or if he ever regretted dabbling into the messy world of biology. After all, a leaf is beyond what a physicist can explain, just as Immanuel Kant said: "There will never be a Newton for a blade of grass." It certainly takes more than a physicist like Qian to explain what happens within a leaf, even though he can build a thing as complex as a missile or a rocket.

The Small World of a Rice Kernel

ONE DAY I brought a few rice kernels to a quiet lab. Sitting in the front of a stereomicroscope, I placed a kernel on the viewing stage. All the fine details of its surface were revealed before my eyes when the kernel was magnified thirty times.

The rice kernel is a microcosm under the scope, with its own topography of hills and valleys. Its husk, the outermost layer, is actually made of two unequal parts, palea and lemma respectively, that are longitudinally joined by an interlocking fold. Two ridges, or nerves, run on the lemma through its long axis. A short, sharp awn extends from the apex of the lemma, and two plumes bracket the base of the kernel. The husk is partitioned to rows and columns of tiny squares, perfectly aligned, creating a pretty and orderly grid pattern. The husk is also bristled with hairs, or spines, which glow and glisten on the illuminated stage because of the presence of a considerable amount of silica in them.

Rice husk is tough, because it consists mostly of fibers and lignin and silica—the rough or abrasive substances that protect the grain from insects and mechanical damages. As rice husk is indigestible to humans, it must be removed before consumption and is often used by farmers for other purposes, such as building material, fertilizer, insulator, fuel, pillow stuffing, etc.

While clumsily trying to tease apart the two halves of the husk under the microscope with a pair of tweezers, I thought of several ingenious ways invented by farmers to remove the husk from the grain. First they thresh the grain to loosen the husk, and then they winnow the grain to remove the dislodged husk. Threshing can be achieved by beating the grain on a hard floor with flail or making mules or oxen walk in endless circles on the grain. Another brilliant idea is to spread the grain on the surface of a country road so that the grain can be threshed by the wheels of passing vehicles. After threshing, the mixture is thrown into the air so that the wind blows away the lighter husk, whereas the heavier grains fall straight to the ground. Farmers had been threshing and winnowing the grain manually for millennia until the development of agricultural machinery such as threshers or winnowing machines. Today, at least in developed areas, a combine harvester can reap, thresh, and winnow the grain

THE STORY OF RICE

simultaneously and effortlessly as it moves along the rows of a field.

Once the husk is removed, a grain of brown rice is revealed. Its outermost brownish layer is called bran, whose main function is to serve as an additional protection against bacteria and molds. Rice bran is highly nutritious, as it contains proteins, oil, vitamins, minerals, dietary fibers, and even antioxidants. However, the presence of an enzyme called lipase in the bran causes rapid deterioration of oil and spoilage of the grain, so the bran is often removed through the milling process as waste and is either discarded or used as animal feed. In earlier days rice was milled mostly by beating the rough rice in a mortar with a wooden pestle. The laborious pounding and rhythmic striking had been part and parcel of every rice farmer's life until milling machines were invented. Today, with an advanced machine, bran can even be removed with husk in a one-step process that produces milled rice directly out of a paddy.

I made a longitudinal section through the long axis of a grain, slicing it as thin as possible with a sharp razor blade. I put the section on a glass slide, added a drop of water to moisturize it, and covered it with a plastic cover. I then put the slide on the stage of a compound microscope designed to view internal structures. The section looked like a mass when it was magnified forty times. I added a tiny drop of iodine. Magically two distinct regions within the grain were revealed, with a sharp boundary. The bulk of the grain turned purplish— the endosperm, in which starch molecules reacted with iodine and changed color. A small portion of the grain in the corner remained whitish; it was the embryo sac containing a miniscule rice plant.

I increased the magnification and zoomed in on the endosperm. It was a mass of tightly packed cells shaped like hexagons, and each of those cells contained aggregates of purple starch granules. The hexagonal pattern reminded me of a honeycomb built by bees to maximize the storage of honey with the least amount of wax and space. Apparently rice has also evolved to use the same geometric rule to economize its resources to store starch in its endosperm, which is the energy source for the initial growth of the embryo once the seed is sowed, somewhat similar in its functionality to a placenta during the development of a mammalian embryo.

Finally I zoomed in on the tiny embryo—the essence of a rice kernel. The embryo comes from a zygote—the fertilized egg. The zygote grows and divides and differentiates, undergoing an orderly series of structural changes and giving rise to an embryo. The embryo itself is a prototype of an adult plant whose future topography is already mapped out. The relative positions of future leaves, stems, and roots are well defined in the tiny embryo, which will unfurl and grow thousands-fold or more once sowed and nurtured.

THE STORY OF RICE

The embryo itself is chemically rich and nutritionally diverse, full of oil, vitamins, and minerals; yet because oil in the embryo tends to become rancid quickly during storage, it is often removed with the bran during the milling process.

After both bran and embryo are removed from the kernel, what is left is a grain of white rice, shining and translucent as a white pearl. The cute dent on it is a vivid reminder that the embryo, once an integral part of the grain, has been removed. Rice grain is a minimalist world. It is small and simple and beautiful. It is chemically pure, gluten-free, and non-allergenic. Although a grain of rice weighs only about a tiny fraction of a gram, the fecundity of rice makes it possible to feed more than one-third of the world's population today and provide a major source of calories to many people.

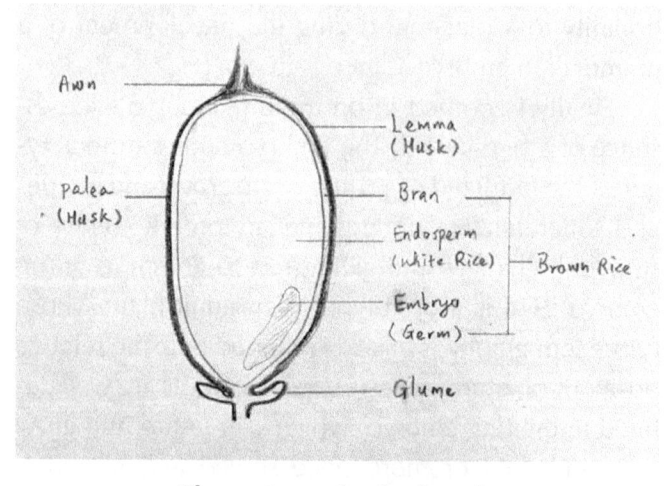

The anatomy of a rice kernel.

The Epic Journey of Carbons

THERE ARE AN infinite number of physical things in the universe, yet there are a total of only 118 elements that make them. Those elements are arranged orderly in the periodic table, based on the number of protons—the subatomic particles—they possess, from one to 118. Not all these elements were created equal—some are rare and scarce, whereas others are ubiquitous and abundant. Some of the common elements include carbon (C), hydrogen (H), oxygen (O), nitrogen (N), and phosphorous (P), just name a few.

Carbon is one of the most important elements, and 95 percent of the things that exist in the universe contain one or more carbon atoms. Carbon has six protons (six electrons as well), and therefore occupies the sixth position in the periodic table. Carbon's six electrons are distributed in the electron shells according to the chemical rule—two electrons in the inner shell and four in the outer shell. But in the atomic world, an atom really wants to have eight electrons in its outer shell to be stable, so a carbon atom often shares its four electrons in the outer shell with other atoms, carbon or otherwise, that also contribute electrons to share, and as a result each of the atoms would have a full outer shell of eight electrons and be in its happiest state. When atoms share their electrons with each other, chemical bonds, or linkages, are created between them to form a molecule.

Carbon atoms can share electrons and bond with each other to form a pure substance, such as graphite and diamond. In graphite, carbon atoms are joined into a honeycomb-like sheet, and many such sheets are then loosely stacked together to form graphite. As the sheets can slide against each other, graphite is soft and brittle, as exemplified by the core of a pencil that is made mostly of graphite. However, in a diamond, each of the carbon atoms is linked to four other carbon atoms, forming a tetrahedral unit, in which five carbon atoms are tightly and rigidly bonded. Hence, depending on the way carbon atoms are joined, graphite flakes away in an instance, but diamonds last forever. Another amazing pure form of carbon is called buckminsterfullerene (or C60), which was named after Buckminster Fuller, who designed the geodesic dome in Montreal. In buckminsterfullerene, carbons are stitched in such a way that the structure looks like a soccer ball or a geodesic dome, at least superficially.

THE EPIC JOURNEY OF CARBONS

The researchers who discovered this amazing structure won a Nobel Prize in Chemistry in 1996.

The real chemical power of carbon, though, is its combinatory power. It is such a friendly and versatile atom that it has an amazing ability to bond not just with like atoms but also with many other types of atoms, such as hydrogen (H) and nitrogen (N) and oxygen (O), resulting in millions and millions of carbon-containing molecules. One of the simplest forms of carbon-containing molecules is carbon dioxide (CO_2), an odorless and colorless gas made of one carbon atom and two oxygen atoms. Carbon dioxide is in the air, in our veins, and in frizzy drinks. It is the gas that makes bread dough rise and makes Champagne corks pop. Its solid form, dry ice, is used as a cooling agent or to generate theatrical smoke or fog on the stage. Despite its ubiquitous existence and versatile usages, carbon dioxide does not contain energy and has no biological uses for us and animals. We breathe it out as metabolic waste.

Another carbon-containing molecule is glucose—a molecule with six carbons and twelve hydrogens and six oxygen atoms ($C_6H_{12}O_6$). Glucose is essential to life. It is an energy-rich molecule that drives biological activities and fuels the growth of living organisms. It is also the building blocks from which other important biological molecules, such as proteins and DNA, are constructed via chemical processes. It is not exaggerating to say that life is powered by glucose.

Although both carbon dioxide and glucose are carbon-based, they represent two entirely different chemical worlds—carbon dioxide is an inorganic molecule and glucose is an organic one, and there is a wide gulf between them. We extract energy from glucose by turning it to carbon dioxide, but not the other way around. Neither animals nor fungi can make glucose out of carbon dioxide; however, some forms of life, such as algae and plants, know how to turn carbon dioxide to glucose by photosynthesis—the very process that transforms inorganic carbons to organic carbons.

Let us travel back in time and trace the origin and history of carbon and carbon-containing molecules. At the very beginning of the universe, there was no carbon or any other atoms, for that matter. A few very elementary and subatomic particles, such as protons, electrons, and neutrons, sprung into existence a few minutes after the "big bang," that thundering moment about 13.8 billion years ago. As the newly created universe rapidly expanded and cooled, simple atoms, such as hydrogen and helium (the first two on the periodic table), formed from those subatomic particles and filled the entire universe. Vast clouds of hydrogen atoms then gathered and condensed, forming the first stars. In the center of those early stars,

THE EPIC JOURNEY OF CARBONS

hydrogens combined to form more complex atoms, such as carbon and oxygen, through a process called nucleosynthesis, or nuclear fusion. When carbon and oxygen met and reacted with each other, carbon dioxide formed, but carbon's journeys into more complex molecules such as glucose were still millions and millions of years away.

The carbons, along with other atoms created in those early stars, were dispersed to the vast universe through supernova explosions. Some of them ended up in a dust cloud called nebula that later formed our solar system—the earth, the sun, and other planets. The earth was born about 4.5 billion years ago from the wandering dust that were pulled together by gravity. However, the earth at its beginning was a hellish place with scorching heat and endless lava. Over the next several millions of years, the earth gradually evolved from a hell to a blue planet, as water was brought to the earth through meteor showers. The water acted as the primordial soup in which small molecules mingled and reacted with each other to form more complex molecules. The first life sprung out of those chemicals about four billion years ago. Since then endless forms of life have evolved on the earth, largely owing to the earth's unique position in the universe—the lucky distance of our planet to the sun makes it a hospitable and comfortable place for our type of life.

Prokaryotes, a primitive type of organisms including bacteria and archaea, were the earliest life forms to exist, and they were single-celled and colorless. The earth stayed as a monochromatic and dull planet until about three billion years ago, when green spots started to glimmer on the top of waters—oceans, ponds, lakes, and streams. The green spots were early cyanobacteria, blue-green algae, that had just invented green pigments called chlorophylls to capture sunlight and use the light energy to make glucose out of carbon dioxide and release oxygen gas as a byproduct—a process we call photosynthesis today. Sometime in the late Proterozoic or the early Cambrian period, another miracle happened; one of those green cyanobacteria had a close encounter with a large cell and was engulfed by it. That cyanobacterium decided to give up its freedom in exchange for warmth and protection, so it stayed within the large cell, acting as a solar panel to harvest light energy and make sugar, and eventually became a part of the cell that we call chloroplast today.

That cyanobacterium-bearing cell became the ancestor of all the photosynthetic organisms on the earth today. Over the eons and eons of time, those specks of green in the ancient waters spread to the wet shores and then to dry lands. While making its epic journey, the green morphed from single-cellular to multi-cellular beings, assuming a variety of shapes and forms—featureless algae, ground-hugging mosses, leafy ferns, bushy grasses, and giant trees that defied gravity. Together they have turned the continents lustrous green, occupying all kinds of habitats, from shores and swamps to mountains and deserts. They contain chloroplasts—the direct descendants of that ancestral cyanobacterium—in their cells. Just like its ancestor, a chloroplast is capable of replicating and multiplying, so a plant cell may contain up to tens or even hundreds of chloroplasts that are busy harvesting light energy and turning carbon dioxide into sugar and releasing oxygen gas to the air.

Photosynthesis is the beginning of the carbon cycle that moves carbons from place to place and from body to body. We are all parts of this elaborate and perpetual cycle. When a plant turns carbons in the air into sugar, a part of sugar molecules is used to build the plant's own body—twigs, bark, or trunks. The plant eventually dies, collapses, and gets covered in earth. After millions and millions of years, its corpse transforms into fossil fuel, such as coal. We burn the fossil fuel for its energy, which releases carbon dioxide back into the atmosphere, completing a carbon cycle. In another scenario, when we and other animals consume sugars or other organic molecules derived from sugars in food, they become our flesh and the flesh of other animals. Carbons are then transferred from animals to us while we are eating them or swapped between animals when they are eating each other. Nevertheless, we all respire and breathe out carbon dioxide as a waste. Whether released from the burning of fossil fuel or from our metabolic activities, carbon dioxide must be brought down again by plants and other green organisms through photosynthesis in order to keep the cycle going. Otherwise, carbon dioxide, one of the notorious greenhouse gases, would trap heat and contribute to global warming, which is happening now as a result of more carbon dioxide released than photosynthetic organisms can possibly recycle them back.

The whole process of photosynthesis occurs in two major stages within a chloroplast. The first stage is called light-dependent reaction, in which sunlight is harvested and transformed into chemical energy. The second stage is called carbon fixation, in which carbon dioxide is turned into sugar with the light energy harvested. Most aspects of the light-dependent reaction of photosynthesis have been conserved during evolution; however, plants have invented a few ways of making sugar from carbon dioxide.

During carbon fixation, carbon dioxide in the air enters the leaf through stomata—the tiny pores in the skin of a leaf—and then it is converted into sugar within the leaf through a series of chemical reactions called the Calvin cycle, named from one of its discoverers, Melvin Calvin.

Most plants fix gaseous carbon dioxide initially into a stable molecule with three carbons; therefore they are called C_3 plants. In a C_3 plant, the initial reaction is a critical and rate-limiting step that is catalyzed by an enzyme called Rubisco (abbreviation of Ribulose-1,5-bisphosphate carboxylase/oxygenase), an ancient enzyme that can be traced all the way back to cyanobacteria and is present in every green leaf and is one of the most abundant enzymes on the planet earth today.

Rubisco is a peculiar enzyme that has dual activities, catalyzing two different chemical reactions. One is to fix carbon dioxide into a three-carbon molecule and eventually to sugar. Another reaction is to fix oxygen gas into a different molecule. The first reaction is desirable, as it leads to the synthesis of sugar, but the second one does not fix any carbons. Actually it is a wasteful process that results in a loss of up to 25 percent of previously fixed carbons. Whether Rubisco catalyzes the first or second reaction depends on the ratio of carbon dioxide to oxygen (CO_2/O_2) in the leaf. A high CO_2/O_2 would promote carbon fixation, and a low CO_2/O_2 would stimulate oxygen fixation. Under normal conditions, stomata are open to allow carbon dioxide in and oxygen out, maintaining a high CO_2/O_2 to ensure that Rubisco fixes carbon dioxide instead of oxygen. However, when the air is hot and dry, stomata are mostly closed to prevent water loss through them. As a result, carbon dioxide cannot get in and oxygen cannot exit, causing a low CO_2/O_2 in the leaf, which leads to the fixation of oxygen instead of carbon dioxide. Plants that grow in the tropical or subtropical areas face a dilemma: if they close stomata to conserve water, they would suffer from the consequence of considerably reduced photosynthesis, up to 40 percent at temperature above 30°C. But if they keep stomata open to let carbon dioxide in, they risk losing water and suffering from drought. During the course of evolution, some tropical and subtropical plants evolved a mechanism of carbon fixation to deal with this dilemma. They invented another enzyme with the sole function of capturing carbon dioxide even when it is low and then concentrating carbon dioxide in the special cells, so that the plants are still able to maintain a high CO_2/O_2 in these cells and undergo carbon fixation even under drought stress. We call this scheme C_4 pathway, because carbon dioxide is initially fixed into a compound with four carbons, in contrast to the C_3 plants that fix carbon dioxide initially into a compound with three carbons. The plants that use C_4 scheme or pathway are, therefore, referred as C_4 plants,

Corn is an example of C_4 plants. Although corn is currently grown in a wide range of climates, it originates from a tropical grass in Southern Mexico. Because corn has inherited the C_4 pathway from its wild ancestor, it is very efficient in recruiting carbons from the air and converting them into sugar molecules, which in turn are made into starch that is stored mostly in its yellow kernels. Because of its C_4

pathway, corn has an amazing high yield, much higher than its botanic cousins that use C_3 pathway, such as rice and wheat, which is why corn has conquered other grain crops and is now dominating the agricultural landscape around the globe. Today corn is being cultivated for many uses because of its high yield. It is grown for human consumption—eaten as corn or processed into cornmeal or corn oil; fermented into whiskey; used as animal feed, turning into the flesh of cows, pigs, chicken, or fish; and manufactured into high-fructose-corn syrup that has insinuated, as a sweetener, into hundreds or even thousands of food products, such as soda, bread, cookies, and ketchup. Furthermore, corn is exploited to make nonfood products, including bio-fuel, plastics, and adhesives. It is no wonder that for a very long time, rice researchers have been impressed by the productivity of corn and pondered if they could turn rice into a corn-like C_4 crop to boost its yield.

It is a daunting task to convert rice into a C_4 crop, however, because the metabolic pathways of rice must be rerouted and its leaf structure must be reconstructed. Many genes, possibly dozens of them, are involved in controlling these biochemical and anatomical traits, and some of these genes are not fully understood yet. Scientists still believe that once these genes are identified and characterized, they can either be introduced into the genome of rice using molecular tools or modified in the genome of rice with

THE STORY OF RICE

gene-editing technology. Currently many researchers around the globe are working together to create C_4 rice. The Bill & Melinda Gates Foundation has provided research funding toward the C_4 Rice Project. It is anticipated that a redesigned rice plant with a complete C_4 pathway would produce more grain—up to 50 percent more—and would be able to do it with fewer resources such as land, water, and fertilizers. Moreover, C_4 pathway would equip rice to better adapt to the hotter and drier environment in the future and continue to feed the growing population. The knowledge gained from rice can be easily transferred and applied to other C_3 crops, such as wheat, potatoes, tomatoes, apples, and soybeans.

Annual or Perennial

RICE AND MANY other grain crops are annuals—they complete their life cycle, from the germination of seeds to the production of new seeds, within one growing season. Rice paddies look sad and grim after the harvest and need to be cleared for the planting of the next season. In regions with a tropical or subtropical climate, sometimes two or three crops of rice are planted each year. Dead stalks have to be removed, the paddy repaired or rebuilt, and fresh seeds sowed twice or thrice a year. Rice farmers' life is busy and hard, and they get no rest from the never-ending and backbreaking labor.

In contrast to an annual, a perennial grass lives for many years. It "hibernates" during the cold winter and grows back in the spring. When it senses the shorter day and chillier air, its leaves and stems start to die, yet its roots and rhizomes in the soil survive through the winter. As rhizomes are actually stems that grow under the ground, they contain growth buds, axillary

THE STORY OF RICE

buds, on their nodes just as on a regular stem. Those growth buds stay dormant in the winter and start to grow at the first sign of spring, sending up new shoots.

Almost daily I drive over a high bridge that overlooks a saltmarsh. The marsh is dominated by several species of perennial grasses. I like to watch the marsh changing its hue as the season progresses. When the chill sets in, the marsh changes from lush green to golden yellow. By the end of fall, the marsh looks brownish—no sign of life is visible except for some marsh birds that occasionally fly over or wade in the water. The marsh always comes back to life in spring, though, shooting up pale-green blades that gradually deepen to emerald by the high summer. The marsh is sustained by its underground world, a seemingly smelly, muddy, and decaying world, yet in it, a vast network of roots and rhizomes holds back, coping with the wintry weather and low oxygen in the soil, processing signals from the environment, and assessing the conditions. When all is right, the roots and rhizomes produce new blades out of their old knobby structures.

Annuals and perennials use different strategies for survival. Annuals such as cultivated rice and wheat put all of their energy and food into seeds to ensure the perpetuation of the species, whereas perennials

ANNUAL OR PERENNIAL

reserve some energy and food in the rhizomes and roots, which provide the fuel and nutrients needed for regrowth in spring. As the cost of sustaining roots and rhizomes is high and is at the expense of sexual reproduction, fewer and smaller seeds are often produced from a perennial grass. For that reason, rice and other cereal crops are annuals, an agronomic trait selected by ancient farmers to maximize the grain yield.

Life is a trade-off. As perennial grasses have to allocate their hard-won calories and nutrients wisely to its seeds, rhizomes, and roots, some scientists believe that a perennial grass cannot have rhizomes and big seed heads at the same time, one of those zero-sum games. Some daring breeders, though, have been trying to defy the rule of trade-offs and develop perennial rice varieties. Some have been crossing the annual rice with the perennial wild rice in the hope that the genes responsible for perenniality and rhizomatousness in wild rice are bred into the annual rice, so their offspring would grow perennially by the rhizomes below and still produce enough seeds to be harvested for human consumption. Other scientists, however, are using a different approach. They focus on identifying and mapping the genes responsible for perenniality and rhizomatousness. Once these genes are characterized and isolated, they can be introduced into the cultivated rice through molecular tools to create a genetically modified rice that will grow perennially.

There might be some potential disadvantages associated with growing perennial rice. For example, it may result in a greater buildup of pathogens and pests in the paddy. It may also make crop rotations more difficult, but there are certainly many good reasons for developing perennial rice or other perennial crops. For example, perennials have a deep, extensive root system. As a result they are more efficient in absorption of water and minerals and are more drought tolerant. They also compete better with weeds, especially annual weeds. They would certainly relieve rice farmers from the backbreaking work of sowing, replanting, weeding, and tilling every season. More importantly, growing perennial rice is a better way to stabilize and improve soil quality, so the land can be used wisely and perpetually. Rice researchers believe that if the perennial versions of today's rice are developed and grown, they could make farming more sustainable and environmentally friendly.

The Death of Grass

PLANTS HAVE MANY natural enemies; every part of a plant, whether a green leaf or a juicy fruit or even a piece of dry bark, is regarded by some of the living things on the earth as a delicious meal. Birds and insects have been nibbling on them and microbes invading them since the beginning of their time. Plants cannot run away from their attackers, so they build physical barriers, such as cell walls and cuticles around their cells. They even make needles and thorns to fend off larger animals. They also hoard nasty chemicals that are either toxic or repellent to their enemies, such as the bitter tannins in an unripe fruit or capsaicin in a hot chili pepper. In the epic wars between plants and their nemesis, both sides have evolved strategies to attack or counterattack, or counter-counterattack—an ever-escalating arms race between them.

The microbes that infect humans or animals are glamorous ones and often make headlines and front pages in the news, such as HIV, West Nile virus, Ebola virus, and most recently COVID-19. However, there are more microbes that invade plants, breaking their defense and causing them to collapse and die. A plant pathogen often spreads rapidly and sometimes leads to an epidemic that could wipe out an entire plant species in a region. One epidemic was Dutch elm disease caused by a fungal species that destroyed large populations of elms in Europe, North America, and New Zealand. Another well-known example was the chestnut blight that occurred in the early twentieth century in North America—billions of magnificent chestnut trees succumbed to a microscopic fungus, resulting in one of the most devastating ecological and economic catastrophes. As chestnut fruits were the major food source for many wild animals, the loss of the chestnut trees resulted in a drastic decrease in the wildlife population in North America. In addition to its ecological impact, the chestnut blight also caused a huge economic loss to the residents of many communities that depended on the trees for food and lumber.

The most devastating plant diseases, though, are those that affect crops—our very sources of food. One horrific plant disease epidemic in our history was the potato blight that occurred in Ireland in 1845, during which a fungus-like microbe infected potato plants and turned them into mushy, brownish masses over a few days. Because potatoes were the staple food for Irish people at the time, the epidemic caused the Great Famine in Ireland and resulted in the deaths of a million and migrations of another million, an event that shaped the history of Ireland and the fate of Irish people.

It is even scarier when a grain crop is infected, because billions of people around the world rely on one or more grains as their staple food. In 1999 a new strain of stem-rust fungus was discovered in Uganda and named Ug99, after the year and country of its discovery. It attacked the wheat crop viciously, causing severe yield losses in Uganda. The fungus spread quickly throughout East Africa and reached places as far as Yemen and Iran within a few years. As more than one billion people on the globe depended on wheat, the rapid spread of Ug99 caused a global concern and even a panic among those who ate wheat-based food daily.

Although a global-wide grain disease has not occurred in recorded history, it is portrayed in *The Death of Grass*, a 1956 post-apocalyptic science fiction novel written by English author John Christopher. In the beginning of the story a new virus has started to infect rice crop in China, causing a major famine and social unrest. The virus spreads rapidly to other Asian countries. As it evolves and mutates, it starts to infect all grass species, including several grains, such as wheat and barley. It eventually reaches Europe and even crosses the Atlantic Ocean to America. As the virus rages on, green pastures turn brown, lush meadows become withered, and livestock perish because of the lack of fodder, followed by the death of birds, fish, insects, and humans.

Luckily the catastrophe depicted in the novel is fictional. Still, a large portion of our crops are lost to various pests and pathogens every year—about a 25 to 41 percent loss was estimated in rice. Why is that? In the past breeders have focused more on developing high-yielding varieties but somewhat overlooked the disease- or pest-resistant traits. Farmers have grown the same high-yielding crop year after year in the same plot—a practice called monoculture farming, leading to the buildup of particular pathogens or pests the crop is susceptible to. Irish potato blight was such a disaster caused by farmers' growing genetically identical potato plants that were defenseless against a particular pathogen.

Rice is particularly vulnerable to microbial pathogens, which are the culprits of many rice diseases, including smuts, blights, spots, and rots. Every part of a rice plant is susceptible to one or more pathogens, and every stage of its growth and reproduction could be interrupted or terminated by them. If left untreated, these diseases can sweep the crop clean and reduce it to a snarled mess.

In addition to being susceptible to microbial infections, rice plants are also often nibbled or devoured by pests such as rats, birds, and armies of insects—aphids, locusts, grasshoppers, weevils, bugs, and beetles. Rice farmers have to use toxic fungicides and pesticides to control these pathogens and pests. For example, parathion had been widely used as an insecticide in the past, and it was condemned as one of the "Elixirs of Death" in Rachel Carson's *Silent Spring*. In her book, Carson documented many parathion-poisoning cases, and some of the victims were small children. DDT, another pesticide Rachel Carson blamed for causing cancer and killing wildlife, had often been used to control pests in rice fields before its ban in the United States in 1972 and later in many other countries.

In an effort to reduce the use of toxic chemicals and minimize their impacts on the environment and human health, scientists have been working on developing new strains of rice or other crops that are immune or tolerant to certain pathogens or pests, either using traditional breeding methods or new genetic engineering technology. On the other side of the arms race, pests and pathogens are also evolving new tricks to overcome the resistance and break the defense of their hosts. It is only a matter of time before a catastrophic pandemic and food shortage will occur if we do not find new resistant genes and breed new varieties to keep up with ever-evolving pests and pathogens. As farmers today tend to grow one or a few high-yielding varieties developed by breeders, we are losing many disease- or pest-resistant genes that reside in traditional varieties and wild species. These valuable genes will soon vanish if we do not save them right now. But how can we save them?

The answer lies in the seeds. Seeds are an amazing reproductive device invented by plants about 350 million years ago as they were crawling gradually from wetlands to parched soils. Since then the embryo of a plant has been encased in a seed and protected by a thick coat. Each embryo is also packed with food for its future growth and equipped with travel gear for its dispersal. Some plants, such as maples and dandelions, furnish their seeds with a pair of wings or a parachute that rides on a breeze to a faraway place. Other plants enclose their seeds in sweet and juicy fruits that attract birds or mammals who digest the flesh and spit or poop the seeds. Some plants even install prickly burs that hitch a ride with a furry animal to a new habitat. A seed may also contain toxins to deter pests or pathogens, like cyanide in an apple seed or ricin in a caster bean. Another trait of seeds we often take for granted is longevity. A seed could last for years, decades, centuries, or even millennia. Its embryo stays dormant and waits for the right moment to sprout. With this trait a seed can be stored for a long period of time, especially under dry, cold conditions. As every seed contains a complete set of genes in its embryo, saving seeds is saving valuable genetic resources, including genes that confer pest- and pathogen-resistant traits.

The good news is that we have started saving seeds. Scientists around the world have been collecting seeds and preserving them in seed banks. There are hundreds of seed banks around the world. The most notable one is the Svalbard Global Seed Vault, which was constructed inside a sandstone mountain on the frozen Norwegian island of Spitsbergen. The area's permafrost keeps the vault below freezing—a perfect condition for seed storage. The vault was also designed to survive catastrophes such as wars and earthquakes. Millions and millions of seeds are safely and securely deposited in the facility, including many varieties of rice. There are also other similar large-scale seed banks, such as the Millennium Seed Bank in England in a nuclear-bomb-proof underground vault that has enough space for the storage of billions of seeds.

In addition to those large seed banks, there are many small-scale and specialized seed banks. For example, the International Rice Genebank, maintained by IRRI, stockpiled more than 130,000 collections of rice, including the cultivated varieties and wild species. It is the largest seed bank for rice and has an incredible amount of genetic resources. A single rice seed in the bank may hold the promise of developing

THE STORY OF RICE

a disease- or pest-resistant variety or a variety that adapts to the warmer climate in the future.

We are saving seeds, one seed at a time. These seeds, secured under dry, cold conditions, will be able to sprout hundreds or even thousands of years later. Breeders and farmers need these seeds, and so do our environment and humanity. These seeds, Jack Harlan, an imminent botanist said, "stand between us and catastrophic starvation on a scale we cannot imagine."

The "Root" of the Matter

ALTHOUGH THE ROOTS of a plant look ugly, they have everything to do with the greenness of its leaves, beauty of its flowers, and deliciousness of its fruits. A plant grows out of the soil just as much as it does out of the air and sunshine, because it gets all its water and minerals from the earth. The uptake of water and minerals is done mostly by the innumerable hairs that emanate from the skin of roots. Although each hair is tiny, there are billions and billions of them. Together they increase the surface area million-fold for absorption. Root hairs are invisible to the naked eye, but you can see the explosive burgeoning of them from the skin of a root on a time-lapse video taken under a microscope. This piece of plant anatomy is often the subject used by photographers for the annual photo-microscopy contest held by the Nikon's Small World—a platform displaying the beauty and intricacy of forms in life through the light microscope.

A plant needs at least seventeen chemical elements

THE STORY OF RICE

in the periodic table to grow and reproduce. Three of those elements—carbon (C), hydrogen (H), and oxygen (O)—are obtained from water (H_2O) in the soil and carbon dioxide (CO_2) in the air. They are then converted to sugar by photosynthesis or incorporated into other compounds through myriad chemical reactions. The other fourteen elements are absorbed from the soil, and they include nitrogen (N), phosphorous (P), potassium (K), and sulfur (S), to name just a few common ones. These fourteen elements play various roles in plants, from making DNA and chlorophyll to regulating metabolism and defending plants against pests and pathogens.

Once the minerals containing these elements are absorbed by root hairs, they pass through several layers of cells transversely toward the center of the root that contains tube-like cells called vessel elements, or tracheids, from which the minerals are transported, along with water, upward to the aboveground parts of the plant. To enter the cell or navigate from cell to cell, minerals need to cross the cell membrane—a thin layer that forms the outer boundary of a living cell. The membrane allows minerals to enter or exit the cell through tiny "doors" embedded in it. These "doors" are called membrane transporters, which are generally very specific for particular minerals. For example, a phosphorous-transporter (Pi-transporter) admits phosphorous only, and a calcium-transporter (Ca-transporter) allows only calcium to go through.

THE "ROOT" OF THE MATTER

Soils, however, are often contaminated with harmful elements that could also insinuate themselves into the root and migrate into the edible parts of a crop, causing health problems to consumers. For example, arsenic (As), one of the natural elements, is often found in the soil. Arsenic is often referred to as the king of poisons, because of its high toxicity and lethality. Even when consumed in small amounts, arsenic can cause cancer, skin lesions, impaired cognitive development, and many other health problems in humans and animals.

Rice is one of the major dietary sources of arsenic, because rice plants are often grown in As-contaminated paddies. Moreover, rice plants tend to absorb arsenic more readily and efficiently from soil than other crops. As a result rice could accumulate an excessive amount of arsenic in its grains and pose a great health risk to the people who eat rice regularly as a staple food.

———∞———

To understand why rice absorbs more arsenic and find a way to reduce the arsenic level in its grains, we have to dig into the roots of a rice plant. The root system of a rice plant, or any grass for that matter, is made of numerous small fibrous roots that radiate outward from the base of the stem and spread among soil particles. When a rice plant absorbs those essential elements needed for its growth, its roots also take up silicon (Si) from the soil that is then transported and incorporated into the upper parts of the plant. Although silicon is not considered one of the essential elements, it helps strengthen rice plants and increase their resistance against insects and pests. Like any other element, silicon in the soil enters the root cells through some special doors called silicon transporters (Si-transporters); yet the Si-transporter has a flaw—it also allows arsenic to slip into the root hair and facilitates its move from cell to cell. Besides Si-transporter, some other transporters, such as Pi-transporter, also admit some arsenic from the soil.

Since the cleanup of rice paddies contaminated with arsenic is not quite a practical and feasible approach, rice researchers have been trying to tackle the problem with genetic methods, creating rice plants that absorb less arsenic. An effort has been made to reduce the uptake of arsenic from the soil by closing the "doors" (for example, Si-transporters) that admit arsenic. Although the closing of these "doors" indeed reduced the absorption of arsenic, it also decreased the uptake of silicon from the soil. As a result those rice plants were weak and vulnerable to diseases and pests. Alternatively researchers have created plants that trick arsenic to exit the vascular tissue or block the upward transport of arsenic into the seed heads. Researchers are hopeful that once they gain more understanding of the genetics of the transport and accumulation of arsenic in rice, they can use molecular tools to reduce the uptake of arsenic and also prevent its upward transport from roots to grains.

In addition to being contaminated by arsenic, rice also tends to be tainted with other toxic elements. For example, the soil in many rice-producing regions around the globe contains high levels of cadmium (Cd), which could be absorbed and accumulated in the grain and lead to kidney failure, bone cancer, and other health problems in humans. Rice researchers have applied molecular tools to reduce the accumulation of cadmium in rice. A research team led by Yuan Longping used a gene-editing tool to knock out the gene for cadmium absorption in rice and therefore bred a variety with a low cadmium content, even when it was grown in cadmium-contaminated soil.

As we know, the soil in many parts of the globe is polluted with toxic metals, including arsenic, cadmium, lead, and mercury. To ensure the safety and cleanness of grains and other crops, we need to both clean up the soil and breed new varieties that absorb fewer toxic elements or accumulate less of them in the edible parts. Molecular approaches, including gene-editing technology, offer some hope of creating crops that can emerge clean and untainted out of the dirty and contaminated soil.

The Book of Rice

SEVERAL ANALOGIES HAVE been used to describe a genome—the total genetic material or DNA of an organism—such as a blueprint, a book, a recipe, or even a movie script, all implying that a genome carries codes or instructions for building the organism and guiding it on its biological journey.

DNA is a long chain of four chemicals called nucleotides abbreviated as A, T, G, and C. An astronomical number of combinations can be concocted from these four common ingredients to generate an infinite diversity of life. A piece of DNA makes up a gene—the genetic unit that codes for a protein of specific function. In a human or animal, the protein may be a hemoglobin for transporting oxygen or an insulin for regulating glucose level. In a plant the protein may be an enzyme for packing starch in the seed, building a wall around the cell, or making green chlorophylls for photosynthesis.

As every aspect of growth and development of an

organism is regulated by one or more genes in its genome, scientists had been dreaming about "reading" DNA since the discovery of the DNA double helix in 1953; that is, to determine the precise sequence of As, Ts, Gs, and Cs in a stretch of DNA. Fred Sanger, a brilliant British scientist, invented a method of "reading" DNA in 1977. With a fine-tuning of his technique and the advent of the DNA sequencer—a machine used to automate the "reading" process—it has become feasible to conduct large-scale DNA sequencing projects. Scientists started to read the books of life in the 1980s. After successful sequencing of several smaller genomes, such as the bacterium *Haemophilus influenzae*, round worm (*Caenorhabditis elegans*), and fruit fly (*Drosophila Melanogaster*), scientists announced the completion of an initial sequencing of the human genome in 2001. In parallel, the genomes of many organisms representing different branches of the "tree of life" were also sequenced around the same time, including the genomes of several plants. Rice was the first crop to have its entire genome sequenced, and the "Book of Rice" was published in a special issue of *Science* magazine in 2002 with a stunning photograph of the Honghe Hani rice terraces in Yunnan province of China on its cover.

The "book of rice" contains 4.3×10^8 letters of As, Ts, Gs, or Cs, organized into twelve chapters called chromosomes. It was estimated that the genome of rice contained about a total of 45,000 to 63,000

THE STORY OF RICE

genes spread throughout its twelve chromosomes. In comparison, a human genome is made of 3×10^9 letters packaged into twenty-three chromosomes, and it was estimated to house about 30,000 to 40,000 genes—fewer than the number in the rice genome—an estimation that surprised many scientists.

Now the rice genome is an open book—all of its chapters are visible and accessible at the digital databases, but it is written in a language we are not quite familiar with. Its grammar and syntax are foreign and incomprehensible to us. Since 2002 rice scientists have been using a variety of molecular tools to study the genome, trying to decipher the hidden biological meanings in it, with a goal of cataloging and annotating all of its genes, one gene at a time.

Through the effort of rice researchers, many genes have been identified and characterized from the genome of rice, including some genes that control the yield and confer disease resistance; yet only a small percentage of rice genes has been thoroughly studied thus far. The majority are still elusive and ill-defined. A daunting task lies ahead for rice scientists to study the biological functions of each of rice's genes, the interactions among these genes, and the interplays of these genes with the environment.

As the technique of DNA sequencing has become cheaper and faster, scientists also sequenced many individual rice plants so they can read the different versions of the rice book and compare them. In 2018 a collection of 3,000 rice genomes representing various varieties or cultivars from eighty-nine countries were completely sequenced through the 3K Rice Genome Project, which uncovered a tremendous number of genetic variations among differing strains of rice. Some of these variations are associated with important agronomic traits, such as pest and disease resistance, nutritional value, yield, or drought tolerance. Rice scientists could harness the valuable genetic resources revealed in these varieties to breed and create rice plants with desirable agronomic traits.

In another aspect, genomic sequencing has transformed the way we study the domestication of animals and plants. The genome of an existing species is a history book, an autobiography, recording the genetic changes of that species over the time and holding the clues to its past. By examining the genomes of subspecies or varieties of a domesticated species and comparing them with the genome of their common ancestor, we are able to trace and delineate the history of domestication from the very beginning. For example, through analyzing the genomes of dogs and their ancestors, wolves, researchers have discovered the changes in the genes that are linked to biosynthesis and processing of adrenaline and noradrenaline, which are involved in the fight-or-flight response. The mutations in these genes have resulted in the diversifications of dog's behavioral and personality traits from their wild ancestors, which is why pet dogs have won our hearts with their docility and playfulness and have become our loyal companions. In the same way, by comparing modern horses with their wild progenitors, scientists identified the changes in the genes underlying the locomotion and musculature, cardiac functions, maneuvering skills, and brain development that led horses to rule the battlefields as well as farm fields.

The same logic also applies to the study of plant domestication. The genome-level comparisons among wild and many varieties of cultivated rice have shed important light on the origin and history of rice domestication. Scientists were able to track down the genetic changes during the domestication of rice and identify the genes that underlined the divergence of cultivated rice. Through such analyses, they were able to trace all the rice varieties of Asian rice around the globe to a single species of wild rice that was domesticated in China thousands of years ago and spread to the other parts of the world, where it was continuously and gradually diverged into many local varieties.

Furthermore, scientists have not only invented how to "read" a genome but also how to "write" and "edit" a genome. Scientists have developed a powerful gene editing tool called CRISPR. Two key players in this innovation—Emmanuelle Charpentier and Jennifer Doudna—won the 2020 Nobel Prize in Chemistry for their contributions. CRISPR acts like a pair of magic scissors that allows scientists to make precise changes in a genome—removing or inserting or substituting a single or more "letters" like a word processor. In humans the technology could be potentially used to remove or edit a faulty gene with remarkable accuracy to abolish an inherited disease. CRISPR, however, has also spurred bioethical concerns, especially for germ-line editing. In November 2018, Jiankui He, a Chinese scientist, announced that his team had edited two human embryos with CRISPR in an attempt to disable the gene coding for a receptor that admits HIV into T cells. Lulu and Nana, twin girls, were born from those two embryos. The news shook the scientific community as well as the general public around the globe, and his work was condemned as unethical and dangerous. He was promptly fired and later sentenced to three years in prison. On the other hand, however, CRISPR has been used by scientists to edit genes in plants with less moral and ethical concerns. So far, significant progress has been made in using CRISPR to create rice with desirable agronomic traits, such as high yield, good nutritional quality, drought tolerance, disease or pest resistance, and low absorption of toxic metals from the soil.

Rice scientists are optimistic and hopeful about the

future of rice, as they are equipped with an increasing understanding of the rice genome, an availability of tremendous amount of genetic resources stored in tens of thousands of rice varieties, and a powerful gene-editing toolkit. They are confident they will be able to create rice varieties that have ideal agronomic traits and also use less of resources such as fertilizers and water, an ultimate goal many rice researchers are striving to achieve in the coming years.

Nature versus Nurture

THE NATURE-VERSUS-NURTURE DEBATE is an age-old philosophical debate that goes back as far as ancient Greece and China more than 2,000 years ago. Plato, a great Greek philosopher during the Classical period, favored nature over nurture and believed that a human acquired traits mostly from his or her parents and the environment played little role, but Aristotle, his student, disputed his idea and believed that a human was mostly shaped by his environment and upbringing. In ancient China, Lao Tzu believed in the natural way, whereas Confucius emphasized the cultural influences on moral development of humans. Such philosophical debates on nature versus nurture have continued, with two polarized views swinging back and forth to this day.

Scientific or biological debates on nature versus nurture did not really start until the publication of Darwin's theory of evolution in 1859. Francis Galton, Darwin's cousin, believed that human traits, both physical and mental, were inherited and not affected by the environment. He was actually the first one to coin the "nature versus nurture" phrase. Inspired by Darwin's theory of natural selection, he proposed that humans could be improved by selective breeding, just as animals and plants had been. Galton launched and championed the eugenics movement in England—the selective mating of the strongest and smartest humans. Such idea of eugenics or biological determinism has been used by many to advance a political agenda or motivate a movement, exemplified by the confinement and forced sterilization of "unfit" people in the United States in the early twentieth century. The ultimate crime was committed by Nazis, who disabled and killed Jews in the name of eugenics. On the other extreme of the debate, the human mind is often analogized as a blank slate that can be written on by upbringing and experiences—an idea supported by Stephen Jay Gould, a famous evolutionary biologist. He said, "The human brain is capable of a full range of behaviors and predisposed to none." Such denial of any possible influence of genes on human growth and development has also made a profound impact on many facets of our society, including criminal justice, educational policy, and even child-rearing.

Today most scientists believe that nature and nurture are equally important—a conclusion drawn from many genetic studies of humans and animals such as bees, rats, and monkeys. To the general public,

NATURE VERSUS NURTURE

however, it seems that plants are irrelevant in the nature-versus-nurture debate, so they have been largely left out of the discussion. The fundamental laws of genetics are universal, however, and applicable to all the living organisms. What is true for animals and humans is likely to be true for plants and vice versa. We may therefore be able to learn a few things from plants by examining the roles of genes and environments in their life.

If you have ever worked in a garden, common sense and experience tell you that you reap what you sow, but you also know that plants must be provided with light, water, nutrients, and warmth so they can grow and develop properly. Apparently both nature and nurture are important to the life of a plant. Luther Burbank, a famous gardener and plant breeder, gained a few insights into child-rearing through his daily work with plants in the garden. In his book *The Training of the Human Plant,* he thought it was time for society to take the next step and create an environment in which children, like plants, could be protected and improved. In his own garden, he nurtured his plants dutifully and helped them become the best that they could be.

―――∞―――

A plant lives a cyclical life, starting from seed germination, going through the stages of growth and

THE STORY OF RICE

flowering, and ending with the production of a new generation of seeds. The progression of the cycle is in sync with the rhythms of the seasons and in response to the environmental cues, such as the change of daylength and temperature. If a plant lives in a temperate region with four distinct seasons, spring would be an ideal time for it to bloom so that it would have at least several months to set seeds and bear fruit before the cold sets in. Because we are so accustomed to the picturesque scenes of spring flowers after a bleak winter, we often take this natural wonder for granted and rarely give a thought to the science behind it. How do these plants know spring is here and it is time to bloom? Actually spring-flowering plants must endure a period of cold before they are ready to bloom; in other words, the plants will not flower unless they go through a few months of wintry weather. In botany this natural phenomenon is termed *vernalization*, coined by Trofim Lysenko, an agronomist of the former Soviet Union. You may ask why plants need to be exposed to cold first. When we think about it, it makes perfect sense, because it is a plant's way to sense the passing of time and seasons. Warm days alone do not necessarily mean spring has arrived, because it could be the result of capricious weather in fall or winter, such as a few unusually warm days. A plant knows that the real spring has come after enduring a prolonged period of frigid cold, so it waits patiently, counting days and yearning for the warmth, as if it knows that "If winter

NATURE VERSUS NURTURE

comes, can spring be far behind?"

Many cereal crops, such as winter wheat and barley, need vernalization. Unlike spring wheat that is sowed in spring and grown into maturity in fall, winter wheat is planted in fall. The young seedlings have to stand in the cold throughout the winter and endue the frost, sleet, and blizzards of the season. If they survive the cold, they bloom in the spring and set seeds in the summer. As winter wheat generally has a higher yield and more proteins in the grain than spring wheat, it is often preferred by farmers in temperate regions, such as North America and Russia.

The story started in the former Soviet Union in the late 1920s and early 1930s, when its agriculture was governed by Stalin's policy of collectivization—the integration of privately owned lands into state-controlled farms. In addition to putting up with Stalin's disastrous policy, the farmers in the Soviet Union also had to face the wintry weathers, which were cold and dry—a harsh and inhospitable condition for growing crops. Sometimes a few extra-cold days in the winter would kill the young seedlings of winter wheat and resulted in crop failure.

In 1928, Lysenko, a young agronomist in the Soviet Union, discovered that chilling of seeds could turn winter wheat to spring wheat, so they could be sowed in spring, evading the chilly winter. He buried seeds in the drift of snow in the winter and sowed them in the spring, and the cold-treated seeds sprouted and bloomed like spring wheat and still maintained its high yield and quality. At the time, the Soviet Union needed his technique of chilling seeds to end crop failure and hunger, so Stalin praised Lysenko as a national hero.

After Lysenko's discovery, however, research and the whole agricultural science in the Soviet Union, for that matter, went off the rails and was distorted greatly by politics and ideology. Lysenko denied the physical existence of genes or chromosomes and opposed Darwin's theory of evolution and Mendel's laws of genetics. He proposed his own theory of inheritance and promoted his own version of genetics. He claimed that plants can be trained, molded, and challenged to perform better and become stronger under adversity. He believed that once the seeds of winter wheat were subjected to cold, the seeds of all future generations wouldn't need any cold treatment and would be able to sprout in spring. He ordered farmers to sow these seeds, which naturally failed to sprout and rotted in the soil. He even boasted that he had turned wheat to rye and warblers to cuckoo birds using his training approach. He promised Stalin that he would transform a barren desert to a green pasture and frozen Siberia to a lush forest in no time.

As his "genetic theory" aligned well with communist ideology and Marxist doctrines, Lysenko's pseudoscience, Lysenkoism, was supported and endorsed by Stalin's regime. As the rest of world moved forward and continued to advance the science of genetics, which eventually led to the discovery of DNA double helix and genetic codes in the 1950s and 1960s, the genetics and agricultural science in the Soviet Union fell into a dark age. Under Stalin's rule the "politically incorrect" scientists who questioned or opposed Lysenko's theory were silenced, exiled, imprisoned, or even executed. In the hands of Lysenko, crops wilted and died in the field; the forest he promised Stalin never materialized in Siberia; the green pasture he imagined was just a pipe dream. His pseudoscience contributed greatly to the famine in the grain-growing areas of the Soviet Union in 1932 and 1933, which resulted in the deaths of three to seven million people.

Lysenkoism also rippled through other communist countries, including Poland, Czechoslovakia, and East Germany, that accepted and promoted Lysenko's theory to varying degrees. However, Lysenko's pseudoscience was completely adopted and Mendelian genetics was utterly denounced in China during Mao's era, because Lysenko's ideas were consistent with those of Mao's thoughts: "man must conquer nature" and "a new and beautiful picture can be drawn on a piece of blank paper." As a direct result of Lysenkoism, agricultural research, especially rice research, in China was greatly affected and much delayed.

According to Lysenko's theory, grafting a branch or shoot of one plant species onto another could create a new plant with novel traits, and the offspring of the grafted plant would also inherit those traits, as well as the following generations. He claimed that the new and better agricultural traits could be generated by grafting alone, and genetic breeding was not needed. Under the influence of his theory, many agricultural researchers in China abandoned their breeding projects and devoted their time and energy to grafting plants, in a hope of producing a hybrid with desirable traits as Lysenko promised.

Yuan Longping was one of those agriculturists in China at the time. He had learned Lysenko's version of genetics during his four-year study at the college. After he graduated in 1953, he applied Lysenko's theory to his work. He kept grafting but yielded no useful plants. After three years of hard and fruitless grafting, Yuan abandoned Lysenko's theory and taught himself Mendelian genetics. He decided to breed plants using Mendel's theory as a scientific guide. After seeing the dead bodies on the street and famished people around him during the Great Chinese Famine (1958 to 1961), he was determined to work on rice, with the hope of creating high-yielding varieties of rice and helping his countrymen. Once Yuan freed himself from Lysenko's pseudoscience, his mind opened and his creativity bloomed. From then on he dedicated his life to rice and eventually succeeded in creating high-yielding varieties in the 1970s.

Looking back in history, both the belief in genetic determinism and the denial of any possible influence of genes had dire consequences. When either of them was used to foster a political agenda or movement, it resulted in disaster and deaths, as exemplified by the horrific genocide committed by the Nazis in the name of eugenics and the great famines that happened in the former Soviet Union and China when Lysenko's environmental determinism was promoted in agriculture.

Fast forward to the twenty-first century. In the age of modern genetics, the mystery of vernalization was finally solved by scientists. A gene involved in vernalization has been identified, and its expression is regulated by temperature. Before exposure to cold weather, this gene is highly expressed, producing a repressor protein that binds to the other flowering genes and switches them off. As a result flowering is prevented. After a period of cold weather or cold treatment, however, this gene is turned off, so it no longer produces the repressor. As a result the flowering genes are on and the plant blooms in spring. After all, vernalization—the botanic phenomenon used by Lysenko to formulate his theory of environmental determinism—is a process orchestrated by both genes and environments, and it is a consequence of the interplay between nature and nurture. Neither genetic nor environmental determinism is right. It is a hard lesson learned from plants.

Extras

IF THE GENOMES of organisms are analogized as books, the sizes of these books have expanded manyfold over billions of years of evolution. The number of the genes—the paragraphs in the book—has increased from just a few to tens of thousands as life evolves from its simplest to its most complex forms.

Looking at a complex genome like ours, scientists found that many of the genes are ancient and are inherited from the older forms of life. As these genes are so essential to life, they have been preserved and passed on from body to body, from species to species, and finally to us. But in our genome or any genome, for that matter, there are genes that are not inherited but have emerged within the species or a population of a species. You may wonder where the new genes come from. Certainly they don't come from nothing, but are they made from scratch with the four basic ingredients—A, T, G, and C—or do they originate from something that already exists? Actually a new gene is often created through a process called duplication—the copying of a preexisting gene. Once the new copy is made, it has its own life. On many occasions the extra copy becomes handy, supplementing or assisting the original one like a spare tire. In other cases, the new copy may mutate and evolve, acquiring a new function along the way. Still, in certain instances, the second copy may degenerate and decay, becoming a relic of the past in the genomic landscape.

Genes are immortal and selfish replicators, as said by Richard Dawkins in his famous book *The Selfish Gene*. Making copies is what a gene is aiming for and good at. A single gene may give rise not just to one new copy, but to tens or hundreds or even thousands of them, through many rounds of duplication over a long period of time. For example, the human genome contains 387 genes coding for smell receptors, which have resulted from many duplication events in the past. Those 387 copies are slightly different from each other because of various mutations accumulated in them over the time. As a result we can smell hundreds or even thousands of different molecules. Still, our sense of smell is dwarfed by rats and mice, which have more than a thousand of genes coding for smell receptors—1,207 copies in rats and 1,035 copies in mice. No wonder that Pixar chose a little rat, Remy, as the master chef in its movie *Ratatouille*, who makes delicious dishes with his gifted olfactory sense. Besides rats and mice, many other animals, such as dogs and cows, are also better sniffers than humans, with a higher number of genes coding for odorant receptors in their genomes (811 copies in dogs and 970 copies in cows).

Although dogs and rats are good at sniffing, they have poor color vision. Their world is bit dull and gray. Both dogs and rats have only two genes coding for opsins, color receptors—one for detecting blue light and the other for green light. They lack an opsin for sensing red light, so they perceive red as a dark shade. Many animals are like dogs and rats with only two-color receptors and are basically colorblind. However, some primates, including Old World monkeys and apes, gained an extra opsin-coding gene through a duplication event, which was then diverged and evolved to detect red light. The extra copy of opsin enables these primates to see reds as well as any colors or shades from mixing of reds with greens or blues, such as purple, orange, or turquoise. As a result these primate species are capable of spotting a ripe fruit in the greenery of a forest or the dimness of wood. As humans evolved from apes, we inherited all three copies of the opsin gene. Thanks to the third opsin gene, our visual world is much richer than that of dogs. We enjoy a ripe fruit not just for its sweetness but also its beauty. We marvel at a rainbow for its seven vibrant hues instead of just two. We admire the artwork created by artists with a rich palette. What would a spring meadow or an autumn deciduous forest look like without reds, oranges, or yellows? Many birds and some insects, however, see more colors than we do, because they have five or six genes coding for color receptors, including the one for sensing ultraviolet and a few others we humans even do not have names for. I wonder what the world looks like to them.

———∞———

Indeed, those extra copies, whether for color receptors or smell receptors, have helped humans or other animals expand their senses and shaped the way they perceive the world today. The duplication of another gene, however, the gene coding for saliva amylase, tells us a different story—a story of how a genome can be altered by diet. Amylase is an enzyme found mainly in saliva that converts starch into simple sugar molecules. If you chew a cracker or a piece of bread long enough, you would get a taste of sweetness in your mouth. It is amylase in action. Most primates, including the extinct Neanderthals and Denisovans, have only two copies of the amylase gene, because starch is only a small part of their diet, and two copies are sufficient for the need. In modern humans, however, the copy number has increased severalfold since a significant number of grains and other starchy foods was incorporated into our diet.

A large variation in the copy number of this amylase gene has been reported among human populations

THE STORY OF RICE

with different traditional diets. Generally speaking populations with a high-starch diet are equipped with more copies of the gene. The high copy number of the gene implies a higher activity of amylase in saliva and more rapid breakdown of the starch in the food. As rice is one of the starchiest foods, it is very likely that rice eaters would have high copies of the amylase gene that help them digest the starch in rice. Indeed, scientists have found that Japanese and other rice eaters, on average, have a higher copy number of the amylase gene compared to populations with a low-starch diet. The varied copy number of amylase gene among human populations tells us that our genome is partly shaped by what we eat.

A similar pattern is also observed in mammals. For example, modern dogs have a higher copy number of the amylase gene than wolves, their wild ancestors. Most wolves have only two copies of the amylase gene, but dogs have four to thirty copies, which are likely to have resulted from their starch-rich diet after they were domesticated by humans. The same trend also plays out in mice and rats, who have lived more and more around humans and eaten more grains and other starchy foods.

The story doesn't end there. It turns out that those extra copies of the amylase gene not only increase our ability to digest starchy food; they also reduce our risk of obesity. Several studies have found a strong correlation between obesity and the copy number of the amylase gene. People with fewer copies of the gene have a much higher chance of developing obesity. Such finding is significant in clinical medicine, as it may provide insight into the prevention or even treatment of obesity. A simple saliva test could be given to a patient and a dietary advice would be offered to the patient based on the level of amylase activity in his or her saliva, so that the treatment plan can be tailored to an individual's starch metabolism and genetic predisposition to obesity.

From One to Billions

MY FIRST CONCEPT of exponential growth came from an Indian legend I heard from one of my second-grade teachers. At the time, I had just learned multiplication tables and could barely comprehend a number bigger than a hundred in real life.

The legend goes: an Indian king was an avid lover of chess and always invited his guests to a game or two. One day the king challenged a wise man who was traveling nearby. The king promised to offer any reward that the wise man asked, if he won. The wise man asked for rice grains that would be placed in the squares of a chessboard in the following pattern: a single grain on the first square, two on the second, four on the third, and eight on the fourth, and the king had to double the number on every consequent one until all of the sixty-four squares were filled. It sounded like a modest request, and the king agreed.

The king lost the game. He kept his promise, ordering a bag of rice to be brought to the chessboard.

He then started to place one rice grain on the first square, two on the second, four on the third, and so on. Before long the king realized that he was unable to fulfill his promise to the wise man. If he kept doubling, he had to put 1,024 grains on the tenth square, 1,048,576 on the twentieth, and more than eighteen quadrillion grains on the sixty-fourth that would be equal to about 210 billion tons of rice—sufficient to cover the entire surface of the earth, oceans and lakes included.

As a child I was captivated by the story. I could connect easily with a fifty-pound bag of rice—the monthly provision for our family of eight. To a child, a fifty-pound bag of rice seemed to contain an infinite number of grains, but I realized that day it was not enough to fill even the first three rows of a chessboard. Later, as a biologist, I have been impressed by the power of exponential growth many times—the tenacious growth of bacteria and the rapid spread of an infectious disease. In 1983 a powerful technique called polymerase chain reaction (PCR) was developed to amplify DNA exponentially and was based on a brilliant idea Kary Mullis, a chemist at Cetus then, conceived while driving on a dark, winding road in California. The technique revolutionized many revenues of science and technology, from medical and forensic science to paleo-biology and ecology. Using the technique, a trace amount of DNA collected from any source, either at a crime scene or an archaeological site, can be amplified to millions or billions of copies in a few hours. Although PCR has never been used successfully to amplify dinosaur DNA as fictionalized in *Jurassic Park*, it has been used successfully to amplify the ancient DNA recovered from mammoth fossils, Egyptian mummies, and human bones. But most importantly, PCR has become a powerful and indispensable tool in medical research and clinical laboratories by its power of doubling and redoubling of DNA. For example, PCR-based test kits have been developed recently to make massive and rapid testing possible for diagnosis of COVID-19 through amplifying the viral genetic material in saliva samples.

Nevertheless, it is the Indian legend that is likely to impress and inspire a young child. I told the story to my older son when he started to learn multiplication, and he was so intrigued by the story that he began to calculate on paper the number of grains that should be placed in each of the squares of a chessboard, but he was overwhelmed by the sheer number and gave up in the middle. Several years later I told my second son the same story, and again the same thing happened. Over the years I have heard different versions of the story and learned that the story has been used by math teachers around the world to introduce the concept of exponential growth. I even found a picture book titled *One Grain of Rice* on Amazon. Because of this story, I don't take a grain of rice for granted anymore and tend to think of rice in terms of numbers.

The number of grains is one of the important indicators for the yield of a rice crop, which is simply expressed by a formula: yield = number of grains x average weight of the grain. It is no wonder that some rice researchers are obsessed with the grain of rice. They have devoted their entire career to studying it, because they know that the grain—both its number and weight—holds the answer to the important question of how to increase the yield of rice to feed the growing population in the future. A slight increase in the number of grains in a single plant can add up and multiply into a significant number in the entire field. A tiny gain in each grain might be the difference between a bumper harvest and a scarcity.

It is very difficult to increase the yield through genetic manipulations of grain number or grain size, though, because both are complex traits controlled by multiple genes that are interacting with each other and with the environment. Some of these genes have been identified, characterized, and even genetically manipulated. However, to rice researchers' great disappointment, they found that grain number and grain weight are interactive agronomic traits that are often negatively correlated. That is, when the alteration of a gene results in an increase in the weight of grain, the number of grains per plant is simultaneously reduced, and vice versa—another example of trade-offs in the biological world. It seems impossible to increase both grain number and grain size concurrently. To resolve this conundrum, rice researchers have to find an optimum somewhere in the middle—an ideal combination of the grain number and grain weight. They can't work alone, though, because it is not just a biological problem but a mathematical one as well.

A complex trait such as the yield of rice is essentially controlled by a network of genes often analogized as an electrical circuit or a computer network. Each gene in the network is installed with a genetic switch that can be turned on or off by a regulatory protein (either an activator or repressor), which itself is the product of another gene that can be switched on or off by another layer of regulatory proteins, and so on and so forth. Furthermore, one network can be interwoven with or embedded in another layer of network—a complicated and mind-boggling system. Just like an electrical circuit or a computer network, though, the gene network still possesses an inner logic or pattern that could be expressed in mathematical language. If all the genes linked to the yield are identified and their interactions with each other and with the environmental factors are elucidated by biologists, a mathematician might be able to write an equation or formulate a model that integrates all the genes and all the switches and all the interactions to predict the yield. The model or equation can then be translated back to biological language and help rice researchers design rice plants through tinkering with genes and their switches in the network, that can reach their highest yield potential in the field.

Mathematics and biology have been intertwined for centuries. William Harvey discovered the circulation of blood in 1628 with the help of simple math—elementary math every school kid can do today. He measured the heartbeat, the volume of the heart, and the amount of blood in the body, and then he did simple arithmetic—adding, subtracting, multiplying, and dividing. Based on his calculations, he argued that the large volume of blood pumped per hour could not possibly come from food consumed and reasoned that blood must be circulated and recirculated within the body. Gregor Mendel relied on elementary probability and statistics to formulate the laws of genetics in the 1860s. Ronald Fisher and others built mathematical models to unite the theory of evolution and Mendelian genetics in the early twentieth century. Math also helped David Ho and its team to model the growth and progression of HIV within the body and figure out that at least three therapeutic drugs were needed to corner the virus, so the cocktail therapy was born out of both biology and math. Today scientists are using math to model the growth and spread of cancer as well as the gene networks that are linked to cancer, so they could use math to guide the cancer treatment and drug therapy. Currently epidemiologists are using math to model the spread of pathogens and help us fight against large-scale infectious diseases such as COVID-19.

If mathematicians can help biologists uncover the hidden pattern of cancer growth and the spread of infectious diseases, they can certainly help rice scientists find the genetic rules and logic behind the yield of rice or any other grain crops. There will be a great opportunity for them to work together to decipher the secret hidden in the grain of rice and find the best solution for maximizing the yield of rice.

Out of Africa

I DIDN'T KNOW wet rice was cultivated in West Africa until I read Alex Haley's novel, *Roots: the Saga of an American Family*. The first part of that book brought me to a haunting landscape and its hardy people in The Gambia during the second half of the eighteenth century. It was a place full of strange yet fascinating things to me, such as the talking drum that relayed news or messages from village to village, the manhood training that transformed boy to man, and the travelers' tree that welcomed visitors or bade farewell to the fellow villagers who were about to embark on a long journey to an unknown place. I was mostly enchanted by a narrative describing how the women in the village went to their rice plots: they rowed in their dugout canoes through the river—one of the tributaries of the Gambia River—to the field each morning during the planting season. Along the way they roused baboons that slept in the depth of mangroves or palm trees, scared wild hogs and made them hide among weeds and bushes covering the muddy banks, and frightened thousands of birds that flapped their wings and took off to the air. The women might even catch a fish or two for an evening dinner during the trip. The description was like a poem or movie clip to me. I read it again and again, conjuring up vivid images of myriad colors and motions.

As depicted in the story, the farmers of The Gambia in the eighteenth century grew their rice in a wet field—the same habitat in which Asian farmers grew theirs, yet the rice grown in The Gambia at the time was African rice *(Oryza glaberrima)*, a distinct rice species that was domesticated independently in West Africa thousands of years ago. Because African rice was not recognized in the original Linnaean classification system established in the eighteenth century, it was not known as a separate species until 1855, when Ernst Gottlieb Steudel, a German botanist, identified it as a unique species and named it *Oryza glaberrima* for its smooth hull (*glaberrima* means hairless or smooth in Latin).

Archaeological studies indicated that African rice was domesticated about 2,000 to 3,000 years ago in the Inland Delta of the Niger River (in present-day Mali) from a wild grass, *Oryza barthii*. The early farmers in that area grew rice on the freshwater floodplains near rivers or creeks. The cultivation of rice spread later to the coastal region in West Africa, extending from Senegal to Liberia—the area that would become known as the Rice Coast, which includes modern-day countries of Senegal, The Gambia, Sierra Leone, Guinea, Guinea Bissau, and Liberia. The farmers in the Rice Coast had grown African rice exclusively until Portuguese introduced Asian rice to the region in the late sixteenth century, but the Asian type was not widely cultivated in Africa until the twentieth century.

Just like Asian rice, African rice is a product of geography and human ingenuity. The farmers in West Africa have developed several systems to grow rice under a wide range of hydrological conditions, from freshwater wetlands to marine estuaries to rain-fed highlands, transforming West Africa into a land of rice agriculture. In the floodplains of several major rivers, such as the Senegal, Gambia, Niger, and Bani rivers, farmers have designed a system that integrates rice cultivation with cattle-grazing. When the tide rolls across the plain and floods the land, farmers grow rice on the wetland. Once the rivers reclaim the floodplains during the fall and winter, the dry land is turned into pasture for cattle, which graze on the rice stubble and debris after rice is harvested, and the cattle's manure fertilizes and replenishes the soil. This seasonal rotation between rice farming and cattle raising is an ingenious and efficient land-use strategy that produces both starchy grain and protein-rich meat for human consumption on the same piece of land.

In the coastal saline estuaries in West Africa, farmers have developed an entirely different system called Mangrove System to grow rice. Unlike the farming practices on freshwater floodplains, Mangrove System requires farmers to clear mangrove forests along the coast and create a rice field in the area. They need to construct an elaborate system of embankments and dikes around the field to gather rainwater and prevent the entry of seawater into the field. Rainfall helps leach the salt from the field once it is built. It takes about two to three years to completely desalinate a field before rice can grow in it. Water management in the field is achieved through a series of canals that control irrigation and drainage. Sluices made from hollowed stems of palm or bamboo and plugged with thatch are used to control water flow into or out of a plot. Development of such a massive system has involved both the ingenuity and grueling labors of the farmers, who have completely transformed the landscape of the coastal area in West Africa to a productive rice field.

In addition to cultivating rice on freshwater floodplains and marine estuaries, West African farmers also grow rain-fed rice in the upland areas that receive more than forty inches of rainfall annually. As the diverse habitats or micro-environments in the upland area present many challenges to rice as well as to growers, farmers have developed strategies to adapt crop to the regional climate and topography from the top of a slope to a low swamp in the upland region.

West African farmers also developed many rice-farming practices and invented farming tools and grain-processing equipment. For example, they invented the transplanting technique independently—sprouting seeds in a nursery first and then transferring the young seedlings to a field. They designed special baskets to winnow grains and also created mortars and pestles to mill them. They adjusted their farming implements for male and female farmers specifically, making hoeing and plowing tools a little easier for them to use.

Another legacy left by West African farmers is the existence of a variety of strains of African rice adapted to different growing conditions, such as deep or shallow water, soil salinity or acidity, drought, and high ion toxicity. There are also varieties that sprout and mature at different times, so that farmers can spread out their labor of planting and harvesting during the busy seasons. There are seeds with long and spiky awns that deter predators and pests in the field. There are also strains that have been selected for the traits related to grain processing and quality, such as ease in milling, storage, and cooking, and desirable taste, flavor, and texture.

If you look at a map and zoom in the coastal area of West Africa and then that of the Low Country (South Carolina and Georgia) of America on the opposite side of the Atlantic Ocean, you may notice an eerie similarity in geology and topology between the two coastlines. The dense salt marsh in the Low Country mirrors the lush mangrove forest in the coastal area of West Africa. Both lands are an amphibian landscape veined by numerous rivers and tidal creeks and strewn with blackwater swamps. In the not-so-distant past, people in both places often had to paddle from place to place in boats and canoes. With a geographic pattern similar to West Africa, coastal South Carolina and Georgia seem to be an ideal place to grow rice, but the Low Country had never seen rice until the late seventeenth century, when rice crossed the wide water and arrived with enslaved Africans. As those slaves were mostly trafficked from West Africa where rice had been grown for thousands of years, they brought their knowledge and skills to the New World and helped plantation owners establish rice cultivation in South Carolina and Georgia.

The slaves in the Low Country converted coastal wetlands, whether snake- and alligator-infested inland swamps or tidal floodplains, to rice fields with an elaborate system of dikes and embankments. The methods and techniques they used were very similar to the West African systems that had been developed and perfected over thousands of years before the slave trade, including the systems developed for rain-fed highlands, inland swamps, or tidal floodplains. The prior knowledge of water control and grain milling from those enslaved Western Africans was essential to the successful cultivation and production of rice, both Asian and African types, in the Low Country, which led to the establishment of rice as a major economy in the area during the eighteenth and nineteenth centuries.

OUT OF AFRICA

After the abolition of slavery, rice plantation owners in the Low Country gradually abandoned their fields and moved away from the area, largely because of the labor shortage and serious damages caused by several major hurricanes. Some of the freed slaves settled in remote villages on a string of the sea islands off the coast, from Carolinas to Florida. Because of their relative isolation while working on the plantations, and later, living on secluded islands for centuries, these Africans formed close-knit communities and developed their unique culture, which came to be known as the Gullah (or Geechee) culture. These people are referred to as Gullah people. The history of the slave trade and rice cultivation of that era has survived through generations of Gullah people, whose legacies are found everywhere on marshy and mossy islands strewn along the Southeast Atlantic Coast.

Gullah language is a hybrid of English with several African dialects. Gullah people eat rice and grow okra. They make sweetgrass baskets and weave strip quilts. They practice hoodoo, as vividly depicted in *Midnight in the Garden of Good and Evil*, a John Berendt novel set in Savannah, Georgia, the Low Country. All those traditions and customs are linked to West Africa and echo the lives of the West Africans brought by force from their native land to the Low Country. The life and story of Gullah people has been portrayed in many books, movies, and music, including *The Water is Wide* (a memoir written by Pat Conroy), *Daughters of*

THE STORY OF RICE

Dust (a movie directed by Julie Dash), and *Porgy and Bess* (an opera by George Gershwin). Johnny Mercer, a great songwriter and native of Savannah, lived on an island inhabited by Gullah people and grew up around them. Mercer found inspiration in the watery landscape of his youth and from the Gullah people he had lived with for years, creating a new type of song with beautiful and enchanting lyrics, including *Lazybones*, *That Old Black Magic*, and *Moon River*. His songs were greatly influenced by the Gullah culture, because he had been listening to Gullah music since the day he was born, when his Gullah nanny sang lullabies and rocked him to sleep.

Many of the Gullah people are the direct descendants of rice-growing people in Sierra Leone—one of the countries on the Rice Coast in West Africa. The Gullah people have reached out to their long-lost families across the sea and made three celebrated "homecomings" to Sierra Leone in 1989, 1997, and 2005, seeking their roots and their ancient connection more than two hundred years after their ancestors were abruptly uprooted from their land. Their heartfelt and tear-jerking trips have been documented in three films—*Family Across the Sea* (1990), *The Language You Cry In* (1998), and *Priscilla's Homecoming* (2005). The story of the Gullah people is currently being told and its culture being preserved through the Gullah/Geechee Cultural Heritage Corridor, a federal National Heritage Area where the Gullah people have been living for centuries, extending from Wilmington, North Carolina, to Jacksonville, Florida.

While learning about the history of African rice and its cultivation in West Africa and then in Americas, I felt that I had found a kindred spirit with West Africans, the spirit born out of the rice field. The picture of a West African woman with a baby on her back hoeing in a rice field in the book *Black Rice* conjured up a similar and familiar image of female Chinese farmers working in a rice paddy. Both Chinese and West Africans express that "a meal is not considered complete unless served with rice." Both regard rice as important in celebrating traditions and performing ceremonies or rituals; both consider rice central to their cultural identity and their homes. People with a history of rice cultivation often form a supportive community, because they have to collaborate, building canals and dikes together, sharing water, and working side by side during the busy seasons of planting and harvest. As a result cooperative thinking and collective actions are very critical for the survival of rice farmers. It literally takes the whole village to grow rice. Indeed, the culture of sharing and collaboration is deeply rooted in both West Africa and many parts of Asia, including China, Korea, and Japan, where rice has been grown over centuries or even millennia.

The world is mysteriously connected. Rice is one of the cultural threads that link Asia and Africa. There have been two parallel paths in the history of rice, each resulting in domestication of a unique type of rice from its respective wild ancestor, establishment of a knowledge system of cultivating wet rice, development of myriad strains that adapt to different ecological niches, and innovations and invention of practices and tools for planting, harvesting, milling, and cooking rice.

Largely because of some of the undesirable traits of African rice, it is being rapidly replaced with its Asian cousin. The seeds of African rice shatter easily in the field, and as a result a significant portion of the yield is lost before the harvest. Its grains are brittle and easy to break if milled by a machine instead of a traditional mortar and pestle. Although West Africans still grow some of the varieties for uses in rituals and ceremonies, many varieties are gone. Sadly, when African rice is disappearing, the good traits or genes associated with it are also disappearing, including its resistance to pathogens and pests in the field and its tolerance to unfavorable growing conditions, such as infertile or toxic soil and drought. It is even more disheartening that we are losing a part of history and heritage associated with the domestication, selection, cultivation, and cookery of African rice.

Rice breeders have been attempting to combine the good traits of African rice and Asian rice; however, it is a challenging task, because the two species cannot be crossed naturally to produce fertile offspring with traditional breeding methods. Recently the rice breeders at the Africa Rice Center in Ivory Coast used a modern technique termed "embryo rescue" and successfully produced fertile offspring after the two species were crossed. Some of the resulting hybrids were found to have the hardiness of the African species and the high yield of the Asian rice. They were also shown to have stronger resistance to pests and diseases. These new hybrid varieties were named New Rice for Africa (NERICA), which holds a great promise in increasing the yield and improving the quality of rice in West Africa, a region still in desperate need of rice as its traditional staple food.

The two paths have crossed now, literally, resulting in a brand-new rice that unites two species domesticated independently by farmers on two different continents. African rice will survive in this new rice, as will its history and cultural legacy.

Bella Ciao

ONE DAY WHILE watching *The Two Popes,* a 2019 movie shown on Netflix, I heard a familiar melody spiral out of the screen that tugged at my heart. Time froze. I paused the movie and searched for the melody buried deep in my brain. Finally I was able to recall the piece of music and realized it was the theme song in *The Bridge*, a 1969 Yugoslavian movie I watched when I was a little girl. I did not remember much about the plot of the movie other than that it was an anti-Nazi war movie, but I did remember listening to the song and humming the tune over and over during my childhood.

The song evoked a wave of my childhood memories, and I got sentimental and decided to learn more about the song. The song is *"Bella Ciao"* ("Goodbye, Beautiful"), an Italian folk song born out of the rice paddy in the Po Valley of Northern Italy around the late nineteenth or early twentieth century. It was the song chanted by rice farmers, especially the women who toiled in the rice paddies. In the song the workers lament about tormenting insects and mosquitoes, wail about their standing in the knee-deep water and bending their backs, protest against their cruel bosses, and yearn for the days when they can work in freedom. Later *"Bella Ciao"* was reinterpreted by the Italian partisans as an anthem of the anti-fascist resistance between 1943 and 1945, inspiring people to fight and die for their freedoms and rights. The song also traveled across the borders and evolved into several versions that had been sung by people of different nations as an anti-Nazi hymn during World War II. The theme song in the movie *The Bridge* was just one of them.

That century-old folk song didn't die off or even get old. More than one hundred years later, the song still lives on. It has become a universal symbol or metaphor for freedom during many historical events, just as it was in the rice paddy in the Po Valley or the battlefield during WWII. It is the theme song in the Spanish television series *Money Heist* as a metaphor for resistance. Tokyo, one of the characters in the show, explained, "The life of El Professor (one of the protagonists) revolved around a single idea: resistance. His grandfather, who had fought against the fascists in Italy, taught him the song, and he taught us." In 2012 the melody of *"Bella Ciao"* was adopted for the song *"Do It Now,"* a song that spurred people to fight against climate change and global warming. The

THE STORY OF RICE

familiar melody and beating rhythm of *"Bella Ciao"* turned into an urgent call for actions to save the planet earth. During the recent pandemic, *"Bella Ciao"* has become an anti-COVID song to encourage and unite people around the world to fight the disease.

Learning about the origin and history of *"Bella Ciao"* was truly a revelation. It took my thoughts to the landscape of the Po Valley, Europe's leading rice producer today, largely because of the abundant water channeled to the area from the Po River and its numerous tributaries. Rice has been grown in the Po Valley since the fifteenth century, perhaps introduced from India. Today rice cultivation in the area is mostly mechanized, from plowing and sowing to spraying and harvesting, but before the mid-twentieth century, the rice paddy in the Po Valley was a scene of desperate poverty and struggles of rice farmers. Everything had to be done by farmers' bare hands. Men were mainly in charge of plowing and maintaining the irrigation canals and dykes, but women did most of the back-breaking and repetitive work in the rice paddy, transplanting, weeding, spreading manure, and harvesting. Because *monda* means *weeding* in Italian, these women came to be known as mondine (weeders). The unbearable working conditions, long hours (from sunrise to sunset), and very low pay led to mondine's deep dissatisfaction and conflict with landowners. They often sang *"Bella Ciao"* or other songs in the paddy to express their sorrows and bitterness. The hardship and suffering endured by these women have been vividly portrayed in *Bitter Rice* and a few other Italian movies.

In the early 1900s, the mondine went on strike to protest against exploitation and demand better working conditions and higher pay. Eventually the women's demands were met and the conflict was mitigated when a mutual agreement on the eight-hour workday was reached. By the mid-twentieth century, rice farmers' working conditions in North Italy had gradually improved and poverty had declined as advanced agricultural machinery and farming techniques developed and agricultural policies, such as land reform and farm subsidies, were implemented.

Today rice grows so well in the Po Valley that it has become the staple food in the region, where rice-based dishes are very common and even more popular than pasta and pizza. Among them, risotto is one of the best-known rice dishes outside of Italy. Risotto is often cooked with short-grain or medium-grain rice and whatever ingredients are in season, from meat and seafood to vegetables and mushrooms, and flavored with cheese, prosciutto, or even cuttlefish ink. Whatever the ingredients are used, they are blended with rice into a flavorful dish by a long, slow cooking technique. Some rice varieties have been specifically developed in the region for making risotto, including Canaroli and Arborio. Risotto, however, is just one of many rice dishes in

Italy, which also include rice soups, rice balls, rice puddings, and many others.

Besides "*Bella Ciao*," there is another well-known song born out of the rice land in a different continent. It is "*Jambalaya on the Bayou*," a song about food, parties, and fun times on the bayou of Louisiana. Unlike "*Bella Ciao*," this song carries a happy, festive tone. In the song, a boy leaves to "go pole the pirogue down the bayous" and attend a party with his girlfriend and enjoy foods such as "jambalaya, crayfish pie, and filé gumbo." Both jambalaya and gumbo are signature dishes in Louisiana. Both are stews cooked with rice and other ingredients, such as fowl, seafood, the holy trinity (onions, celery, and green bell peppers), or whatever is in season and available locally.

A pot of jambalaya or gumbo is truly a melting pot, just as Louisiana itself. It is a product of multiple cultures and cuisines that have been blended harmoniously together into a single delicious dish. Jambalaya or gumbo echoes, but not repeats, the flavors and cooking styles of several rice-based dishes originated in different parts of the world, such as Spanish paella and Italian risotto. Several key ingredients of different geographic origins have found their ways in the same pot. For example, okra—a thickener used in making gumbo because of its slimy texture—was

BELLA CIAO

brought to America by the slaves from West Africa. It was speculated that the gumbo dish was named after okra's name—*ki ngombo* or *quingombo*—in the Niger-Congo languages spoken by many enslaved Africans. Filé powder—another thickening agent—was invented by Choctaw Indians from dried sassafras leaves. As the Choctaw word for filé is *kombo*, some people argued that gumbo might have gotten its name from the Choctaw word. Roux, another ingredient often used to make gumbo, is made from fat and flour, a practice that has its origin in French cooking.

Recently I ran into an interesting blog about the Hitachi rice cookers in Acadiana, a rice-growing region in Louisiana. Lucius Fontenot, a photographer and native of Louisiana, took photos of old Hitachi cookers many families in Acadiana still owned or used. Those old Hitachi cookers chimed when rice was ready, serving as a dinner bell in old times. In his photo gallery, some of the rice cookers looked battered and worn after thousands of rice dinners had been cooked in them, but others still looked shiny and new. Behind each of those rice cookers there was a heartfelt or entertaining story. Some of the cookers had been with the family for three or four generations and were passed down as family heirlooms. One family lost its Hitachi rice cooker during Hurricane Katrina and later found a similar one to replace it for sentimental reasons. An elder woman got hers as a wedding gift, and it outlasted her marriage. According to the blog, Louisiana was the place that had the largest number of Hitachi rice cookers in North America in the 1960s, which beguiled the Hitachi company in Japan, so it sent a representative to Southwest Louisiana to investigate the curious phenomenon. The man understood why when he saw miles and miles of rice land as if he were standing in his native country Japan. Reading this story was another revelation for me: when Louisianans started to use electronic cookers to make their rice, the people in China, the closest neighbor of Japan, was still using firewood and coal to steam their rice and continued to do so until the early 1990s.

Just as the Po Valley, Acadiana of Louisiana is another example of the places where food is intertwined with its geography and culture. Like the Po Valley, Southern Louisiana is a perfect terrain for growing rice, largely because of its abundant supply of water and subtropical climate. As rice made its way from the Carolinas through Georgia, Alabama, and Mississippi to Southern Louisiana in the nineteenth century, it soon replaced wheat and corn crops that had been cultivated but ill-adapted to Louisiana's wet climate. Since then rice has been a major cash crop for farmers in many areas of Southern Louisiana. Crowley, a city in the heart of Louisiana rice country and nicknamed "Rice Capital of America," has hosted the International Rice Festival since 1937. It draws hundreds of thousands of visitors each year who participate and enjoy rice-themed activities, including a Rice and Creole Cookery Contest, Rice Eating Contest, and the Coronation of the International Rice Queen.

New Iberia, another city in Louisiana's rice country, boasts of having the oldest rice mill in the United States—the Conrad Rice Mill. Established in 1912, it is listed on the National Register of Historic Places. The mill is still standing and operating today as a witness to the history of rice cultivation in Louisiana over the last hundred years.

The Stairways to Heaven

GUILIN HAS ALWAYS been a mythical place to me. It is a scenic city in southern China, a land of hard rocks and fluid water. The sepia-toned image on the back of the Chinese 20-yuan banknote captures the essence of Guilin's geography, with jagged karst peaks mirrored in the placid Li River, a lone cormorant fisherman paddling a raft in the water, and groves of bamboo standing tall on the riverbanks.

In her 1989 debut novel *The Joy Luck Club*, Amy Tan takes her readers to Guilin where the Joy Luck Club is founded. She starts her story by waxing poetic on the beautiful hills of Guilin. At the time Guilin was still a safe zone in the midst of Japan's invasion of China during World War II and flooded with hungry and ragged refugees. But Japanese soon started to air-raid Guilin after many cities of China fell. As sirens warned people of incoming bombers, residents and refugees naturally ran to the craggy caves underneath those cone-shaped hills and hid for days like cave

THE STAIRWAYS TO HEAVEN

animals. Those natural bomb shelters, some as big as chambers, saved many lives during war times.

Guilin's mountains and hills are renowned not because of their height or size, but their oddity and quaintness, rising from the plain abruptly, almost at right angles. This mountainous land was once a deep ocean in which ancient fish and crustaceans swam along with long-extinct marine reptiles such as ichthyosaurus and nothosaurus. On its floor limestone was formed from the shells and skeletal fragments of those ancient marine animals. When the momentous collision of India with Asia lifted the iconic Himalayas about 50 million years ago, the same crash force also raised the limestone-rich ocean floor in the area. The exposed rocks were then sculpted and carved by nature's hands—the soft part washed away by acid rains or chiseled off by winds over the eons and eons of time. The hardest rock remains to this day, forming literally a "forest of karst peaks" in Guilin.

On a summer day my husband and I took our two children to Guilin to catch up on our shared childhood dream of seeing this fairyland. With its subtropical climate, Guilin was hot and humid in the summer. I got hot and sweaty, but once on a boat and cruising down the Li River, I felt a cool breeze and was invigorated and refreshed. Those frozen images on the banknote or Chinese paintings came alive, becoming three dimensional and vivid. Seen from the slow-moving boat, the landscape was revealed like a sideways scroll painting unfolding before the eyes. Every section of that painting was picturesque and miraculous. The mountains and hills were transfigured before my eyes into elephant noses, horse heads, or giant fishes. The layered peaks at the far distance, vague and bluish, changed their tunes and profiles, moving in and out of sight magically.

Travelers are invariably mesmerized by the natural beauty of Guilin's karst peaks, yet Guilin is also famous for its other mountains—the mountains carved by human hands and blanketed by a single plant species—rice. If you happen to look at the Google satellite map of Guilin, you may notice an intriguing image in its landscape. Viewed from above are the regions snaked by hundreds of thousands of curved lines that stretch for miles upon miles. They are the Longji rice terraces and the home of Yao and Zhuang people—two of the fifty-five ethnic minorities in China. *Longji* means "dragon's spine" in Chinese. Those spines are actually a series of flat surfaces, like tiered steps, constructed into the slopes of the mountains. They flow gracefully and fluidly, following the natural contour of the rolling land.

We took a tour bus to the Longji terraces, a two-hour ride from the center of Guilin city. Bustling streets soon gave way to rolling countryside. As soon

THE STAIRWAYS TO HEAVEN

as we got off the bus, we started to climb to the summit, along a narrow and muddy road and under an on-and-off mountain rain. Dark gray clouds hung over the terraces; the drizzling rain obscured our view. However, the weather didn't prevent us from appreciating the scenery. The mountain under a veil of mist looked mysterious and dreamy, like a fairyland. Under our feet raindrops rippled the water in the paddy, and the rice seedlings danced in the breeze.

When we reached the top, the rain had stopped and dark clouds dissipated. The air after the rain was refreshing and transparent. The terraces stretched as far as the eye could see. Nothing could compare with what I saw at that moment. It was a world of flowing lines and green waves. Layers upon layers of paddies looked like stairways, the water-filled pools of various sizes rising one above another were like glass shards shattered from a giant mirror and then pieced together, glittering and shimmering under the afternoon sun. The farm houses that had been concealed earlier were now revealed, nestled in the hills and enveloped in greenery. At dusk we stood and watched the sun setting. The terraces, like Monet's garden, acquired new characters and charms with each passing minute. We stood transfixed until the last ray of sunlight slipped away and the terraces went to sleep.

The Longji Rice Terraces.

We continued to walk on the winding and slippery road the next day. The hotel's small dog decided to follow us for a while. We chatted with fellow travelers and villagers and were often passed by fast-walking local women wearing colorful handmade traditional costumes and carrying baskets on their backs. They supplemented their income by delivering food and supplies to the hotels on the hill or carrying travelers' luggage to their rooms.

After a long walk we rested near a stream, drinking the icy-cold spring water gushing out of the mountain. I saw a villager washing and plucking a chicken on the side of the stream, imagining that the chicken would be soon chopped into pieces and stuffed into the hollow of a bamboo stalk and roasted over a fire—a famous local dish called bamboo chicken. We stopped at a farm house and met its residents—a sixtyish woman with an infant boy sleeping on her back and a seven- or eight-year-old girl by her side. The girl was exceptionally beautiful with her sparkling eyes and rosy cheeks. We chatted with the woman for a while and learned that the parents of the two kids had left the village, joining many other migrant workers in the factories or construction sites in the city and leaving the kids behind in the care of their grandparents. Along the way a middle-aged couple invited us into their tall stilt house. We bought a freshly picked cucumber from their garden and a bowl of soy custard made from soybeans with an old-fashioned millstone. While eating the cucumber and soy custard, we watched the woman weaving a piece of beautiful fabric on an old loom while several chickens searched for food and left their droppings around the house.

My mind keeps drifting back to Guilin every now and then, to those magical karst peaks and the mystic Li River, but I thought more about the terraces and the people who built them. William Blake once said, "Great things are done when men and mountains meet." Indeed, Zhuang and Yao people have created a marvel, sculpting the rough, hard mountains into picturesque and breathtaking stairways. I also thought about the little girl and boy. Will they stay to continue the legacy created by their ancestors? For a tourist like me, seeing the terraces from a distance, all was beauty, harmony, and pastoral idyll. To a farmer who works daily and closely in the terraces, though, he sees mud, swarming mosquitoes, and endless back-breaking work with little monetary gain. Because it is impossible to use modern machinery on a terraced field, all the work has to be done by human hands with a little help from draft animals such as buffalos or cattle. As modern ways of living are spreading to every corner of the globe, it seems inevitable that young people will eventually leave the terraces and abandon the home and community built by their forebears. In the future the sublime landscape of the Longji rice terraces will probably be seen only in photos or films. Alas, just like many other traditions, the Longji rice terraces will eventually become a nostalgic story and a cultural relic.

Kill Two-birds with One Stone

RICE IS A unique crop. Unlike other grains that are grown on dry land, rice is cultivated mostly semi-aquatically. Its entire root system and lower part of the shoot are submerged in water throughout the growing season. Other grains would suffocate in water because of the lack of oxygen, but rice thrives.

Although rice could grow on dry land, it does much better and yields more in a flooded paddy for good reasons. Flooding prevents the growth of weeds that compete with rice for nutrients and sunlight and space. Flooding also wards off rats and other pests that feed on rice. Because of these benefits, rice is cultivated mostly in water-filled paddies around the globe.

Rice can flourish in an oxygen-deprived paddy because it has numerous air spaces (aerenchyma) in its roots and culms (stems), so that the air breathed by its aerial part can be circulated and sent down through these air spaces to its lower region immersed in the water. The air chambers in rice are an anatomical legacy left by its ancestors that grew in swampy habitats. Similarly, many other aquatic or semi-aquatic plants, such as water lilies or lotuses, have also evolved to have air spaces in their roots or stems for breathing. A trained botanist is often able to tell whether a plant is aquatic or semi-aquatic by looking for the unique anatomy associated with aquatic living, such as a spongy and porous tissue that can hold air. One of my favorite aquatic plants is the lotus, which has graceful flowers that decorate ponds and lakes around the world and also has a porous root (it is actually an underground stem known as a rhizome), with six to nine large air chambers arranged in a beautiful pattern in its cross section. A dish made of slices of lotus roots is delicious, healthy, and also pleasing to the eye with its aesthetic beauty shaped by living in the watery world.

Just like plants, many animals have also been built by evolution to live semi-aquatically, such as frogs and ducks who spend at least a part of their time in water. Instead of porous and spongy tissues, a different set of anatomical features have evolved in semi-aquatic animals. For example, a duck is covered with oil-coated feathers for waterproofing and equipped

KILL TWO-BIRDS WITH ONE STONE

with a pair of webbed feet for paddling. A duck is also kept warm and cozy in frigid water by a layer of soft down under its feathers and another layer of fatty tissue beneath its skin. A duck looks graceful and poised in a pond, stretching its curvy neck and moving with its webbed feet, but it looks clumsy and slow on land, ambling along with a heavy buttock and a pair of short legs. Ducks appear to enjoy being in water more, swimming and diving with ease. A friend of mine who raises ducks on her small farm told me that it took her and her husband quite a bit of time and effort every evening to get their dozens of ducks out of the pond and into their coops to sleep.

As food, ducks are popular in China, and they have been cooked many ways in China since ancient times. Their eggs are often salted or pickled. A perfectly pickled duck egg has a deep orange-hued, buttery yolk that goes well with rice porridge, a popular breakfast food for many Chinese. Even duck feet are turned into a delectable dish in China. Beijing duck (or Peking duck), a roasted duck, however, is the most famous duck dish in China and best known outside of China. It is coveted by many people for its characteristic tender meat and crispy skin. When wrapped in a thin crepe together with green onion and sweet bean sauce, it is a heavenly dish. Beijing duck and other roasted ducks are often served at Chinese restaurants overseas. Hanging in a window with their heads intact, those fat-dripping ducks look enticing and appealing

THE STORY OF RICE

to a Chinese person but can be horrifying to others. In the movie *A Christmas Story*, the Parker family has its turkey ravaged by dogs at the last minute, so they go to a Chinese restaurant for their Christmas dinner, where they are served a roasted duck complete with its head. The family members are shocked and amazed when a "smiling" duck is brought to the table and horrified when it is decapitated on the spot. It is a highlight of *A Christmas Story* and the most memorable scene for many audiences. In the movie the duck is aptly called Chinese turkey, as many oversea Asian families often substitute turkey with duck when preparing their Thanksgiving dinners. Duck meat is much juicier and tastier than turkey meat, thanks to the fat tissue that keeps ducks comfortably warm in cold water, yet instead of marbling through the meat like beef, fat in a duck is layered between meat and skin, so duck meat is fairly lean once the fat layer is stripped off.

As both rice and ducks thrive in a similar habitat, some rice farmers have devised an ingenious way of raising them together so they can harvest two products from the same plot, which means more income to them. When reared together, rice and ducks benefit from each other tremendously. Ducks eat insects, snails, and weeds in the paddy, and they also enrich the soil with their droppings. As a result less

KILL TWO-BIRDS WITH ONE STONE

fertilizer, pesticide, and herbicide is needed in the paddy. Ducks swimming freely and eating live insects rather than being caged and fed with manufactured food also lead a much happier life than those raised on the factory farm. Such a dual-farming system produces much better rice and heathier ducks.

If ducks can be raised in a rice paddy, logically other aquatic or semi-aquatic animals could also be grown together with rice. Indeed, Chinese have raised fish in their rice paddies for at least a thousand years. Just like rice and ducks, rice and fish benefit from each other in the same way—fish eats insects and weeds in the paddy and rice is fertilized by fish poop. One of such rice-fish farms in China was designated one of the globally important agricultural heritages by the Food and Agricultural Organization of the United Nations in 2005. The farm has been developed by generations of farmers in Qingtian of the Zhejiang province. Since its recognition, the village has attracted many tourists to see the unique rural landscape and eat fish and rice harvested from the rice-fish paddy.

Similar dual-farming systems have also been used to raise shellfish such as shrimp, crabs, and crawfish in rice paddies. In Louisiana farmers have been raising crawfish with rice for many years and developed a very efficient system and farming routine. In March or April, farmers plant rice and flood the field. In May or June they add crawfish brood stock into the water. In August the field is drained and rice is harvested, during which the crawfish burrow underground and survive in the wet mud. After the harvest, the pond is re-flooded and crawfish feed on rice detritus left in the field. Crawfish that have grown to market size are harvested continuously from November until April when the field is drained again. Rice is then replanted and the new growing cycle starts. Such rice-crawfish dual farming is so profitable that rice breeders even developed a rice variety known as *ecrevisse* (a French word for crawfish) in the early 2000s, specifically for raising crawfish.

Such a duel-farming system of rice with an aquatic or a semi-aquatic food animal is a sustainable, productive, symbiotic, and eco-friendly way of farming. It also brings economic benefit to farmers, who harvest two things from the same piece of land, truly killing two birds with one stone.

Naming Matters

EVERYTHING DESERVES A name, whether it is a living thing or an object. Without a name, the history of its existence would vanish. Everything also belongs and has special relationships with others in the world, which is why naturalists and scientists have been obsessed with naming things and organizing them into groups and subgroups. Once a thing is named and put into a proper place, we are able to talk about it and see how it is connected to the rest of the world.

Naturalists and philosophers had been trying to name and organize life as far back as ancient Greece; however, before the time of Carl Linnaeus (1707 – 1778), a Swedish botanist and physician, the animals and plants on our planet were not properly named and categorized. As a result, pigs and elephants might have been lumped together based on superficial resemblances; so might palms and pines. Sometimes the same species was given two different names or different species the same name. Herbaria were chaotic and cluttered places back then, lacking a universal organizing scheme.

Carl Linnaeus devised an elegant system that we still use today. In his system, species are ordered in a hierarchical manner based on their morphological and anatomical features, starting with kingdoms. Each kingdom is divided into classes, each class into orders, each order into genera (singular: genus), finally, each genus is divided into individual species. Linnaeus's approach displays the complex relationships among organisms in a simple tree-like hierarchical structure.

The Linnaean system also gives each species a scientific (Latin) name that consists of two words. The first word indicates the name of genus the species belongs to, and the second word is the specific epithet for that species. The first word is analogous to our last name, reflecting the species' ancestry, linage, and belonging. The second word is unique and idiosyncratic, and it can be named after a person, a place, or anything the namer desires. For example, the scientific name for modern human beings is *Homo sapiens*. *Homo*, which means human being in Latin, is the genus that encompasses modern humans and several extinct species, including Neanderthals and Denisovans, whereas *sapiens* means *wise* in Latin. In another example, *Agra schwarzeneggeri*, a species of beetle, was named after Arnold Schwarzenegger by a fun-loving scientist. The name carries a hint of masculinity because the male of this beetle has well-developed biceps-like middle femora. Still in another case, *Amblyomma americanum*, the Lone Star tick, was named after America, the country where it is commonly found.

The Linnaean naming scheme sometimes also gives names to taxonomic units below the rank of species, such as varieties, cultivars, or subspecies. An animal or plant may therefore have a three-part name, indicating its genus, species, and subspecies (or variety) respectively. For example, the scientific name of purple finches is *Carpodacus purpureus*, whose two subspecies are named *Carpodacus purpureus ssp. purpureus* (Eastern purple finch) and *Carpodacus purpureus ssp. californicus* (Pacific purple finch) respectively.

Linnaeus was a master organizer and sorter in the world of biology. He transformed a chaotic jumble of life forms into a neat, organized world in which species are linked by their natural similarities. Since his time the system has been used by scientists around the world and in different time periods, so the knowledge built for a particular species can be perpetuated and shared by all.

The Linnaean system did not merely tidy up the herbaria and galleries in a museum; it also transformed the way scientists thought of the natural world and the relationships among different forms of life. Charles Darwin, while working on his theory of evolution, was influenced by Linnaeus's way of thinking and inspired by the order created by the Linnaean system, but unlike Linnaeus, Darwin believed that the

THE STORY OF RICE

order was created by natural selection and not by the mind of God.

———∞———

The scientific name of Asian rice is *Oryza sativa*, and it was named by no other than Linnaeus. *Oryza* is a Latin word for rice and *sativa* means *cultivated*. Asian rice contains two major subspecies—short-grained and long-grained. The scientific names of two subspecies were coined by a Japanese scientist in 1930—*Oryza sativa* ssp. *japonica* for short-grained and *Oryza sativa* ssp. *indica* for long-grained. Obviously the names of these two subspecies are Latin words for Japanese and India. Although such naming of two subspecies of rice has been adopted by the scientific community, it has been greatly disputed and protested in the past by Chinese rice researchers who thought such naming disregarded the history of rice. China, the origin of rice domestication and cultivation, was left out, as if irrelevant and unimportant.

Chinese rice researchers argued that the two types of rice had long been recognized by ancient Chinese, and their descriptions were recorded in the ancient texts 2,000 years old. In Chinese the two types of rice were given separate names—粳 (geng) for short-grained and 籼 (xian) for long-grained. Ding Ying (1888 – 1964), the pioneer of rice research in China, proposed the name change of two subspecies from *japonica* to geng and from *indica* to xian, but it was difficult to change a name that was internationally accepted and standardized. Besides, Chinese rice researchers were less visible than their Japanese counterparts at the time and didn't have much to say in the matter of naming rice.

Fast forward to the present day, and China has caught up with the rest of the world in science and technology and even taken the lead in the area of rice research. Japan and China competed fiercely with each other in the sequencing of rice genomes at the beginning of the twenty-first century, with China focusing on xian and Japan on geng. Chinese out-competed Japanese and published the entire sequence of xian first in 2002, ahead of Japanese team. In 2018, Chinese completed another daunting task—sequencing 3,000 rice cultivars. This time, when the scientists submitted their work for publication in *Nature* magazine, the authors used xian and geng in their article and even inserted the two Chinese characters 籼 and 粳 in the text. The editor thought that the use of Chinese names was against the internationally established standard and asked the authors to remove them, but the authors didn't budge and with their passionate pleading and citing of the history and the ancient texts finally convinced the editor to accept the names. It was perhaps the first article in English language journals or any other non-Chinese language journals, for that matter, that contained Chinese characters, but it certainly was not the last one.

Cell, another prestigious science magazine, published a paper on the genetic mechanism that conferred chilling tolerance in *japonica* (geng) rice in one of its 2015 issues. On its cover is an art form of 田, a Chinese character representing a paddy field, in which a cartoonish farmer is transplanting rice seedlings in the knee-deep water of a paddy. It is a vivid and delightful image. Later, on the April 2019 issue of *Plant Biotechnology Journal*, four abstract-looking 米, a Chinese character for rice, appeared on its cover, glistening against a dark background. Each of four 米 is also encircled with a perfect ring, forming a Zen-like image. Looking carefully, one will find that those four 米 and the rings around them are assembled from grains of rice. Each of the six strokes of 米 is a string of rice grains linked together. Looking more carefully, you will find that four 米 are different sizes, even though they are pieced together with exactly the same number of grains. Why so? Because they were constructed with the grains harvested from four different rice plants—one of them regular and the other three genetically manipulated plants. As explained in the article published in that issue, the grain size of rice could be significantly enlarged through genetic manipulations of a single gene in rice. The cover is an artistic rendering of the authors' important discovery. The image conveys not only a scientific message, but also a cultural one.

NAMING MATTERS

The magazine covers with artistic forms of Chinese characters. "田" is a Chinese character representing a paddy field; "米" is a Chinese character for rice, and four of them are pieced together with rice grains. (Courtesy of *Cell and Plant Biotechnology Journal*)

North and South

THE CONCEPT OF north-south divide extends far beyond geography; it often refers to a political, cultural, or even socio-economic divide. North and South Koreas separated by the thirty-eighth parallel are ruled by different political regimes. The Mason–Dixon Line symbolizes a cultural divide between the North and the South of the United States, and a bloody civil war had been fought between them. Such divide, either real or imagined, has also been perceived in the history of several countries, including the United Kingdom, Vietnam, Italy, and other countries.

In China, the north-south schism has been recorded since ancient times. Geographically China's north-south boundary is roughly defined by the Yangtze River that runs eastward from the highland of Tibet to the East China Sea, dividing China into two parts—the north and the south. Many episodes of the north-south conflict occurred in the history of China, including several epic wars and famous battles waged between

NORTH AND SOUTH

them. Some of the politicians and military leaders and diplomats involved in those conflicts at the times have become legendary figures. The events that defined the victories or defeats have been dramatized and romanticized in folklore, movies, and literature. The Northern and Southern Dynasties (420 – 589) of China, aptly named, was a period when a high level of polarization between the north and south was reached, during which two sides referred to each other as barbarians.

Northern and Southern China vary in their climate and geography. The north is generally cold and dry, consisting largely of flat plains and grasslands and deserts, whereas the south is warm and wet, with rolling hills veined with rivers and streams. As a result rice, a crop adapted to a hot and humid climate, is grown mostly in the south; whereas wheat and millet are cultivated mainly in the cool, dry areas in the north. The Yangtze essentially splits China's agricultural landscape into rice and wheat regions and divides its population roughly into rice-eating and wheat-eating people. Each has developed a unique cuisine around its staple food.

——∞——

Westerners, however, had often lumped together the different groups of Chinese and regarded them as one, mostly because of the history of Chinese immigration. Between the mid-1800s and early 1900s, a large number of Chinese emigrated to the United States and other countries, driven largely by the poverty, wars, natural disasters, and unemployment during the late Qing dynasty. It was the first big exodus ever in the history of China. Those immigrants were mostly from Guangdong (Canton), a province on the southern coast of China. Attracted by the hope of finding gold in America, they crossed the Pacific Ocean and settled in San Francisco, which they dubbed Gold Mountain. Just as any group of immigrants, they clung to their old dietary habits and ate rice and Cantonese-style dishes. In the eyes of the Americans of 1800s and early 1900s with European ancestry, Chinese were seen as people who ate rice with two wooden sticks, wore a long-braided queue, and spoke choppy and comical-sounding English. Some of these stereotypes have persisted into the twenty-first century as generations of children grew up reading books filled with jarring racial stereotypes, including Dr. Seuss's *To Think that I Saw it on Mulberry Street*, in which a Chinese person is depicted with a pair of chopsticks, a rice bowl, a pointed hat, yellow skin, and slanted slit eyes.

Although the image of "rice-eating Chinese" is widespread throughout the media or movies with scenes of Chinese dinner tables or Chinese take-outs, in actuality, people in northern China eat more wheat-based foods, such as noodles, pancakes, buns, or dumplings. I had personally experienced the dietary dichotomy between northern and southern China during my growing-up years, as I spent several years with my paternal grandparents in Shanghai, a city in the south, before I reunited with my parents and three siblings in the far north of China, where my maternal grandparents also lived with us. It wasn't difficult for me to adjust to the dietary change from rice to wheat after I moved from south to north at five years old, but my father came to the north from Shanghai in his twenties, married my mother, settled down, and raised four kids. He never got used to northern food and always yearned for the food he ate in Shanghai, as if his stomach was already molded by rice and southern-style food. I witnessed the same struggle in my college classmates who came from the south and ate in the cafeteria that served mostly wheat-based food at the time.

Besides the dietary difference, there are some psychological and behavioral dissimilarities between the northerners and southerners in China, either real or assumed. Northerners are often considered to be outgoing, rough, and generous, whereas southerners are viewed to be prudent, introverted, and shrewd. Some of these regional stereotypes have been deeply ingrained into the minds of many Chinese people, affecting their thinking and social interactions with each other. The regional stereotypes were even promoted by some of the prominent people in China.

THE STORY OF RICE

One was the Kangxi emperor (1654 – 1722), who was the fourth emperor of the Qing dynasty and often considered one of China's greatest emperors, because of his many military and economic and literacy achievements. Among 246 pieces of advice he gave to his children, he said, "The people of the North are strong; they must not copy the fancy diets of the Southerners, who are physically frail, live in a different environment, and have different stomachs and bowels." Lu Xun, one of the greatest writers in modern Chinese literature, wrote, "According to my observation, Northerners are sincere and honest; Southerners are skilled and quick-minded. These are their respective virtues. Yet sincerity and honesty lead to stupidity, whereas skillfulness and quick-mindedness lead to duplicity."

The psychological and behavioral differences between the northerners and southerners in China have been featured in movies, depicted in essays and literature, and joked about by comedians but had never been investigated in a scientific way until 2014, when several American psychologists collaborated with Chinese counterparts to conduct a large-scale study. They published their findings in *Science* magazine, arguing that the psychological and behavioral differences between the northerners and southerners in

China were attributed mostly to the farming practices in the two regions. As rice farming is very labor intensive and requires a significant amount of water, rice farmers need to cooperate with each other to build a complex system of irrigation. Sometimes they have to negotiate and compromise about their access to water. When rice ripens, farmers often need to work together to harvest it as quickly as possible so they can start another season right away. Because the survival and success of rice farmers depend largely on cooperation among members of the community, they are more interdependent and reliant on each other, as a result forming a close-knit community and collectivistic culture. In contrast, wheat fields are managed quite differently. Wheat farmers do not have to depend on each other as much. They do not need to build canals and dikes together. They generally farm only one season, followed by a long period of idleness. Consequently wheat farmers are more individualistic and analytical thinking.

To eliminate other confounding variables that might have also contributed to the north-south difference, such as climate, cultural heritage, or historical events, the authors of the paper compared the neighboring counties along the rice-wheat border, because the difference, if any, between them would be less likely to be caused by climate or other variables. Again they presented convincing evidence that people from the rice side of the border thought more holistically than those from the wheat side.

Another intriguing aspect of their study is that none of their thousands of participants had actually grown rice or wheat for a living themselves. They just happened to be the descendants of rice or wheat farmers, yet the psychological and behavioral traits were still passed on and persisted into the current generations long after people put down their plows and abandoned farming.

Nobody knows exactly how the traits associated with farming practices are transmitted, whether through cultural inheritance or biological mechanisms, and how far these traits will persist into future generations. As a geneticist, I can't help relating these interesting and intriguing questions to a revolutionary concept of modern genetics called epigenetics. According to epigenetics, our genes are being tagged constantly by special chemical groups throughout our lives, in response to our experiences and environments, social or physical. Some of these chemical tags can switch genes on and others switch them off, therefore influencing or altering our genetic traits. Moreover, these chemical tags accumulated throughout a person's life can be passed down to children, grandchildren, and beyond. In other words, the egg from your mother or the sperm from your father contains not only the genes that make you, but also some chemical marks that have been acquired as a result of your parent's life experiences. These chemical marks are able to turn your genes on or off and predispose you to certain genetic traits, including psychological or behavioral attributes.

A seminal study was conducted on the survivors of the Dutch Famine (or the Dutch Hunger Winter), a famine that took place in the German-occupied Netherlands during the winter of 1944 – 45, near the end of World War II. A German blockade cut off food and fuel supplies to several provinces in the Netherlands, and as many as 22,000 people died and about 4.5 million were severely affected and barely survived the famine. The study showed that the children who were born to the starving men and women during the Dutch Hunger Winter were more predisposed to anxiety, depression, and many other physical and psychological conditions; yet the most surprising result came from a study on the grandchildren of men and women exposed to the famine. They too had higher rates of those physical and psychological conditions, suggesting that some heritable factors or chemical tags must had been imprinted into the genomes of starving men and women and passed down across at least two generations.

In addition to this groundbreaking discovery, epigenetic phenomena have also been widely observed in many other studies on humans and other organisms. If our ancestors' experiences of hunger or other vicissitudes can be passed down to us and become a part of who we are, it is logical to think that our ancestor's farming practices in the rice field may also influence us in a similar way, predisposing us to work hard, cooperate with each other, and think holistically. We still need more scientific studies, though, to determine if epigenetic mechanisms are indeed responsible for the association of certain psychological or behavioral traits with farming practices.

In *Outliers*, Malcolm Gladwell makes a connection between rice farming and high level of mathematical competence of Chinese and other Asians. He argues that the work ethic and mathematical skills of Chinese and other rice-farming Asians have been shaped in part by growing rice, which demands hardworking, precision, multitasking, and careful planning. I agree with the author that the "rice paddy makes a difference in the classroom, and growing rice makes you better at math." The current generations of Chinese are truly the beneficiaries of the legacy left by their rice-farming ancestors. The belief system and logical thinking skills shaped by thousands of years of rice farming have been passed down to them through cultural transmission or even by biological mechanisms such as epigenetic inheritance, propelling them to work hard in math and preparing them for the intellectual and technological challenges in the twenty-first century.

The math talent unveiled in the new generations of Chinese had been hidden and untapped in previous generations. When Albert Einstein toured China in the 1920s, he was appalled by the "industrious, filthy, obtuse people" he saw in China. He wrote in his diary, "The Chinese are incapable of being trained to think logically, and they specifically have no talent for mathematics." He even lamented about the possibility that "the Chinese may well supplant every nation through their diligence, frugality, and abundance of offspring." He continued, "It would be a pity if these Chinese supplant all other races. For the likes of us, the mere thought is unspeakably dreary." Although a great thinker, an insightful philosopher, and an avid advocate for social justice in the public eye, Einstein plainly expressed his prejudice and racist views on Chinese and other ethnic groups in his private thoughts. He failed to realize that the Chinese people of the early twentieth century were mostly poverty stricken and education deprived, and they never had an opportunity to develop their innate mathematical ability and logical-thinking skills, another example of how nature and nurture are intertwined and inseparable in life.

Most recently the culture of collective responsibility and cooperation has been demonstrated in Asians during the pandemic. People in China and some other Asian countries are willing to give up their individual rights in exchange for collective safety, and they don't mind sacrificing their freedom for the greater good. Even the Asian communities overseas are complying willingly with mask mandates and shelter-in-place orders. Asian Americans currently have the highest vaccination rates in the United States and other countries. Cultural heritage really matters in a crisis like this one.

Reap What You Sow

WHENEVER I RIDE a train in China nowadays, I always make sure I get a window seat so I can look out the window and enjoy the scenery. Once the train leaves the station and passes through the ever-increasing high-rises of the city, I gaze with pleasure at the open landscapes—rolling hills, meandering rivers, and boundless farmlands. In southern China the countryside is often dominated by rice paddies, which are picturesque with chessboard-like geometry and glimmering water under a blue sky or with luxurious growth in green, gold, or amber, depending on the season. Sometimes in a gust of wind the whole field ripples like ocean waves—a spectacular scene I will never forget.

My husband was born and raised in one of the rice countries in China—Zhejiang, an eastern coastal province that borders Shanghai to its north. Zhejiang literally means zigzag river in Chinese and refers to the Qiantang River, an important commercial artery in

THE STORY OF RICE

China. The river runs for 285 miles through the province, snaking through its capital, Hangzhou, before joining the East China Sea.

Zhejiang is known as the "land of rice and fish" in China, one of the most commercialized and richest provinces in China, owing partly to its geographic advantages and rich cultural heritage. But my husband did not grow up in Hangzhou, a city known as "a heaven on the earth" or any other prosperous city with a glorious history. He grew up on a farm near Huangyan, a small coastal town. For generations his family had grown rice and other crops on the farm. He started to work when he was only four or five years old, first as a helper doing small jobs in the paddy, such as pulling weeds and gathering scattered grains, and then as an adult when he reached fourteen. He did all kinds of hard work, from transplanting and weeding to spraying and harvesting. As a boy my husband hated farming and dreamed of living a different life—the life he had glimpsed in the few movies he had watched. He was so elated when he was accepted to college and was determined not to go back to being a farmer again.

He has been off the farm for more than forty years, now, and the time has sweetened his memories and softened his hard feelings. Now he brags to his sons about carrying a nearly one-hundred-pound load over a hill when he was ten. He talks, with a tone of bittersweetness, about picking up the bagasse freshly spitted out by sugarcane chewers on the street and using it as fuel. He reminisces about swinging from tree to tree in an orange grove, catching eels hidden in the mud of a rice paddy, feeding a goat with a small frog or dragonfly wrapped in a wad of grass and laughing when the goat spat it out, or playing poker games with fellow herders on a hill while goats or cattle leisurely grazed nearby. He claims proudly that he cooked family dinners on an open-fire stove since he was only seven. It is as if his memories have been filtered by passing time and only the fun parts have remained. Those stories made me envious of his childhood and regret that I didn't grow up on a farm and have nothing to brag about. But he is not forgetful about the hardship and suffering his parents and grandparents experienced and endured. With misty eyes and a choked voice he often talks about his mother collecting or even begging for fresh orange peels door-to-door and then drying them and selling them as a gradient for Chinese medicine to earn a little money for his and his brothers' school fees.

I met his maternal grandparents several times before they passed away, and both lived to a ripe old age and farmed until the end. They were the epitome of rice farmers of their time in China, with a sun-scorched skin, weathered faces, and stick-like frames clad in an oversized shirt and baggy pants. They were the most unpretentious, frugal, and practical people I had met in my life. His grandfather was born in the year of the 1911 Xinhai Revolution—the revolution that overthrew the Qing dynasty and led to the establishment of the Republic of China in 1912 and brought the thousands of years of imperial ruling to an end. His grandma arrived just a few years later after the birth of the Republic of China. During the late Qing dynasty, China was ravaged by imperial corruptions, internal rebellions, natural disasters, and wars with Japan and European countries. It was a land of famine and chaos throughout the nineteenth century. Dr. Sun Yat-sen, the founding father of the Republic of China, said, "What is considered inequality between poor and rich in China is a difference between the very poor and less poor," and he promised the poverty-ridden peasants that "those who till the land should have the land."

Dr. Sun's promise never materialized, though, after the establishment of the Republic of China. Instead the thirty-seven years of the republic were filled with constant political and military upheavals, including seventeen years of warlordism, eight years of Japanese invasion, and three years of civil war. China was wracked by bloody battles and natural disasters. Its economy collapsed and its people were devastated by food shortages. Chinese peasants suffered most from the high taxes levied on them, the forced seizure of their crops by armed men and bandits, and endless natural disasters and crop failures. The hardship endued by the Chinese peasants in that era was depicted in *The Good Earth* written by Pearl Buck, who lived in China then, mostly in the Zhejiang province, and witnessed the sufferings of Chinese peasants firsthand.

As a young man, my husband's grandpa toiled in the rice paddy during the busy seasons and worked as a porter during the slack seasons, carrying goods between towns to bring in a little money. It was usually a two-day journey with a one-hundred-pound load on both ends of a bamboo pole balanced on his bony shoulders. In one of his long, backbreaking trips during the Chinese Civil War, he ran into a group of young soldiers of the Chinese People's Liberation Army—the armed force of the Communist Party—while they were fighting against the Nationalist Army. A young soldier told him, "You will not have to work this hard once we win the war." Like many poor peasants at the time, he welcomed the new China wholeheartedly with anticipation and hope.

Following the victory of the Communist Party in 1949, land was taken away from landlords and redistributed among peasants. Initially the poor peasants were pleased to own a private plot, yet the dreams and hopes of those poor peasants were shattered once again by the ever-shifting agricultural policies and manmade disasters during Mao's era. In 1952 the government began to organize the peasants into collective farms called people's communes, owned and operated entirely by the government, with the intention of increasing the agricultural output and raising currency to finance China's industrialization. The farming inefficiency created by this system, however, led to low productivity and food shortage. Furthermore, in 1958, Mao and his party launched the Great Leap Forward campaign, with an ambitious and unrealistic goal of catching up with the United States, the United Kingdom, and other economic powers within a few years.

During the three years (1958 to 19961) of the Great Leap Forward movement, Mao and the party adopted many agricultural policies and farming practices from the Soviet Union that had failed catastrophically two decades earlier in its own country. Mao repeated the same mistakes made by Joseph Stalin as if he was not aware of the mass starvation and food shortage that happened in the Soviet Union in 1932 and 1933. Chinese peasants were asked to plant seeds and seedlings close together, a practice promoted by Lysenko, a Russian agriculturist who believed individual plants within the same species never competed with each other but helped each other, like individual people in the same social class. Naturally plants withered and died from overcrowding. Peasants were also asked to plough deep, at least five-feet deep, again a practice used widely in the Soviet Union. As a result seeds rotted in the deep soil and never had a chance to sprout. Peasants were further ordered to make fertilizers—again an approach of the Soviet Union—by mixing manure and household rubbish, such as bits and pieces of glass, bricks, and mud. These absurd and laughable practices turned many fields into wastelands. Furthermore, many farmers were pulled out of their fields and ordered to make iron and steel from their household items or kitchen utensils—pans, pots, or knives—to accelerate China's industrialization, but instead of making usable iron and steel, the peasants made poor-quality and useless junk with their crude backyard furnaces—a huge waste of labor and resources.

In the midst of the madness, even little sparrows became an enemy of the whole nation. Mao listed them as evil, because they ate rice and other grains. Under his order the whole country was commanded to make loud noises with drums, pans, or whatever available to scare sparrows and prevent them from landing on trees or roofs. The poor sparrows flew and circled endlessly in the sky until they dropped dead from exhaustion. Once sparrows disappeared, insects multiplied and locusts swarmed because of the lack of natural predators. These pests stripped crops clean. Worst of all, in an intention of improving irrigation, peasants were called to work day and night, constructing dams, canals, and reservoirs, yet they were designed by peasants instead of engineers and built with mud instead of concrete. As a result most collapsed within a few years—dams reduced to rubble, canals abandoned, reservoirs filled with anything but water. In 1975 the Banqiao dam built in that era in Henan province burst after a torrential storm, releasing a gigantic volume of water that drowned at least 26,000 people.

In 1959 instead of verdant growth in the spring and golden harvest in the fall, many farmlands in China were bare and bleak, yet local officials were fearful of Mao and the party, and rivaled each other to fulfill the quotas based on the exaggerated yield projection, taking away the last grain, literally, from the peasants and leaving them to starve. Hungry peasants dug up roots, peeled off tree bark, foraged for wild greens, and collected grass seeds to fill their empty bellies. By the end of fall, when birds flapped their wings to go south for warmth and food, there was nothing left for the peasants to eat. They huddled in their sheds cold, hungry, and wasting away. Some people gathered their last strength, kneeling, wailing, and begging for food on the streets or at front doors. Roadsides were scattered with bony bodies and corpses during the year of 1959. In some cases people resorted to cannibalism when their morality and ethical values decomposed along with their bodies and senses. The horrific famine repeated year after year until 1961—three years in a row. The Great Leap Forward resulted in tens of millions of deaths. Most of them were rural peasants.

Alarmed by the widespread famine in China, some pragmatic senior party members realized the mistakes and criticized Mao's ideology and policies. Under pressure from some of his colleagues, Mao relented and agreed to implement new policies, allowing peasants to raise livestock and grow food on their private plots, increasing the state's purchasing price for grains, and reopening the markets and trades. Under such policies, the large-scale famine was mitigated by the end of 1962, although peasants were still hungry and poor. During the next few years, Chinese peasants were able to breathe a bit easier and feed their families, but in 1966, Mao returned to the political arena with a vengeance and regained power. He launched the Cultural Revolution, a movement that targeted and persecuted those who criticized his Great Leap Forward movement and his disastrous agricultural policies. During the Cultural Revolution that lasted from 1966 until Mao's death in 1976, he tried to revive communes and revert to the same policies that had caused the Great Chinese Famine, leading again to many episodes of crop failure and poverty.

Chinese peasants had to wait until the early 1980s to get their own plots and gain some freedom after Mao's death in 1976. The communes were gradually dissolved and market economy was instigated by the new leader, Deng Xiaoping. Inspired by Deng's famous saying, "The cat that catches mice is a good cat, whether it is black or white" and spurred by his slogan "Getting rich is glorious," Chinese peasants entered a new era—an era in which they could listen to the wisdom of their forebears and their own, and they started to see the hope for a better life.

In 1990 we visited the village where my husband's grandparents still lived in. While strolling on the country road and chatting with people in the village, we were under the impression that the peasants were still very poor and struggling with legacies left from the Great Leap Forward and the Cultural Revolution—the land ruined by deep ploughing and irrigation projects poorly designed and built. The village had no running water or electricity, and peasants still used their hands or draft animals to pull ploughs. To my husband the place looked the same; nothing had changed much. He pointed to the hill where he had herded goats, the mountain trail he had trodden with heavy loads, and the rice field where he had caught eels or carp. When we went back just a few years later, however, it was a different world. New multistory houses made of brick and installed with electricity and tap water were being erected everywhere. Farming was mechanized with the arrival of tractors and threshing machines. More and more households owned electronics and appliances such as televisions and microwaves. Although my husband's grandparents still lived in the same mud hut, they owned a piece of land and made a little money by renting it out to young farmers.

REAP WHAT YOU SOW

———∞———

Farming has always been hard, particularly during the pre-machinery and pre-fertilizer era. The same struggles and heroic efforts have been repeated and have echoed on every piece of land and in every corner of the world, whether a plot of prairie land worked by pioneers, a mountainous slope cultivated by Peruvians, parched soil plowed by Africans, or even a small bean field hoed by Henry Thoreau near Walden Pond. Farmers have always been mud bound and lived in the margins, but Chinese peasants who lived in Mao's era suffered the most and endured one hardship after another, living in perpetual poverty and under oppression for more than a quarter of a century.

From that snapshot of history, we learned that the land has its own natural laws and logic. It does not simply yield to man's will and power. It has to be loved and nurtured by the people who own it and have an intimate relationship with it. Mao and his party believed that man should conquer the nature and declared a war against nature and land. Under his rule mountains were cast into terraces or valleys to grow crops; forests were reduced to stumps to fuel steel factories; and all enemies, even sparrows, were killed to clear the way to the victory of humans. His philosophy and approaches ended in economic and ecological disasters in China. Today the same land

THE STORY OF RICE

that had been fruitless and bare during Mao's time yields much more when cultivated by sensible farmers and facilitated by new crop varieties, fertilizers, and machinery.

We met a young rice farmer in his thirties several years ago in my husband's hometown and toured his rice field. He, in a clean shirt, khaki pants, and a pair of brown leather shoes, proudly showed us his three hundred acres. It was a new era for a new generation of rice farmers. Unlike his parents and his grandparents, who used their bare hands and raw labor, this young farmer, empowered by science and technology, grew rice on a scale that his parents and grandparents could not even imagine.

My in-laws visited us a couple of times, helping us take care of our infant sons. We had a small vegetable garden then. Sometimes we grew Asian greens in the gardens, such as yau choy, a Chinese version of cabbage; cucumbers with tiny pricks on the skin; and eggplants with fruits shaped like bananas, but we never had any luck with Chinese celery, which has thin, flexible stalks. We craved the strong aroma and medicine-like flavor of Chinese celery, but it was difficult to grow. Even after we coerced seeds to sprout with all of our tender and loving care, the seedlings withered and died for no reason. We blamed our failure on

REAP WHAT YOU SOW

our sandy soil that did not hold water well. When my in-laws came along, they took over the garden. My father-in-law had the greenest thumb I had ever seen. In no time the garden burst with all kinds of greens; among them was a stand of tall Chinese celery, with which we delightedly made dumplings and stir fries. My father-in-law was a quiet man and rarely uttered a word during his visit, but he showed his feelings and communicated his love to us in that garden; that tiny patch of earth was his stage, on which he expressed himself.

Recently I watched *Minari*, a 2020 film depicting the struggle and isolation of a family of South Korean immigrants in rural Arkansas. Minari is also called water celery and is an ingredient in many Korean or other Asian dishes. In the movie the grandma takes her grandson to a stream, searching for a patch of land to sow the seeds. In the end, when everything else fails, that patch of celery is growing well and flourishing, symbolizing the resilience and adaptability of immigrants. Minari in the movie looks very much like the celery my in-laws planted in our garden many years ago, but unlike minari, the celery we grew was a type of dry celery, and it was more vulnerable and needed a lot of care. Nevertheless, whether adaptable or susceptible, these celeries have brought familiar smells and tastes to immigrants and connected them to their homeland.

Rice farmers like my parents-in-law have a deep

THE STORY OF RICE

connection to the land and an intuitive understanding of the rhythm of seasons and the cycle of growth from their daily toil and sweat in the rice field. They take notice of all the nuances of soil, water, light, and temperature. They truly understand the meaning of "reap what you sow" literally and metaphorically. In addition to sowing and reaping their crops, they also sow the seeds of discipline and self-reliance and work ethic in their children. Those seeds have taken roots and grown in their sons and daughters, giving them a solid foundation and the ability to face life's challenges. Although many of their children have left the farm and are pursuing other careers, they take the culture of the rice paddy to whatever the jobs and tasks they do, whether a difficult math problem, a hard project, or a physical exertion. They, in turn, are passing the culture of the rice paddy to their own children and beyond, which is how cultural heritage endures through time.

Go to Space

PLANTS OFTEN ASTONISH us with their amazing mutability and adaptability. They seem to be able to figure out how to fit into any environment they are brought into. Although both natural and artificial selection have a lot to do with the diversity and wide distribution of plants, it is the plants' genetic richness and willingness to change that make it possible for them to survive and thrive in a variety of habitats, from arctic tundra to deserts and from high mountains to low plains. They obligingly take on a new form or color we and nature throw at them. Who would be able to tell, by a glance, that cabbage, broccoli, cauliflower, kohlrabi, and kale are all the same thing—a wild mustard (*Brassica oleracea*) in disguise? Who would not be amazed by the fact that *cannabis* has reincarnated into two plants with different chemistry—cellulosic hemp and psychedelic marijuana—without knowing a bit of history of their domestication?

Plants spread as humans spread. When people

migrate, they always take seeds with them in their pockets or sacks. Since ancient times seeds have walked trails; climbed hills; ridden horses; boarded trains, ships, or airplanes; and settled in faraway places with people, which was how rice got out the swamp of southern China and ended up in almost every corner of the world. Because of its vast genetic repository, rice has been able to reinvent itself again and again with help of humans during the past 10,000 years, conquering one territory after another and making itself at home in places as different from one another as Japan, Sierra Leone, southern Louisiana, and the Po Valley. Rice's tens of thousands of genes have shuffled and reshuffled millions of times, resulting in about 50,000 genetically distinct varieties today, each of which has a unique combination of traits and adapts to a particular climate and geographical region.

As humans are naturally born travelers, they will never cease moving and exploring. It is inevitable that humans will eventually go beyond the small blue planet and venture out into space—the moon, Mars, or some other earth-like planets, either to search for unknowns or to save ourselves from catastrophic events such as an incoming asteroid or a manmade disaster.

The question is this: where are we going first? Mars or the moon or others? Although the moon is much closer—a three-day trip—Mars is thought the most livable planet in our solar system by far, because it has ice water, a higher gravity compared to the moon, and a bit of atmosphere. Mars is still a hostile place for our species, though, with its absence of oxygen in the atmosphere, lack of liquid water on the surface, extreme coldness and dryness, and high level of radiation. We must therefore find out how to survive there before we can actually colonize that red planet. While other space scientists and engineers are working on building machines and devices that could convert ice to liquid water and extract oxygen from carbon dioxide on Mars, botanists are busy figuring out how to grow plants in space.

Food is one of the most important necessities for humans to survive in space. In the sci-fi movie *The Martian*, astronaut Mark Watney awakens after a terrible dust storm on Mars and finds himself injured and left behind by the crew, and his immediate concern is food. Being the mission botanist, he knows a great deal about plants. He grows potato plants from the crew's leftover potatoes using Martian soil and the crew's feces as fertilizer. He also manufactures water from leftover rocket fuel to irrigate his potato garden. I was so involved in the movie that I breathed a sigh of relief when I saw green sprigs shooting up on the screen. Mark Watney survives on Mars for more than a year, largely because of his makeshift potato plot. Imagine what would happen if another astronaut who knows nothing about gardening were to be left behind. He or she wouldn't know that every tiny eye on the skin of a potato is a growth bud that has the potential to sprout into a new plant and that a dozen potatoes could be transformed into a sizable garden. The movie is compelling and inspiring, and it left me with a thought-provoking question: can we really grow plants on Mars?

On the earth a plant grows in response to gravity, sending its roots downward (a phenomenon termed positive gravitropism) and its shoots upward (termed negative gravitropism), which is how a plant orients itself and senses its place so it can maximize the exposure of its upper parts to sunlight and ensure the close contact of its roots to the minerals and water in the soil. It had long been thought that the major challenge in space gardening was to get plants to grow in micro or even zero gravity. However, it has been shown by astro-botanists that plants could be tricked to grow in space; gravity was not absolutely necessary for the growth of plants, because a plant could orient itself by responding to other environmental stimuli such as light (phototropism) and touch (thigmotropism). In 1982 the crew of the Soviet *Salyut 7* space station was able to grow lettuce successfully in space, and the lettuce plants were able to flower and set seeds. Since then many plant species have been grown in space, including some economically important crops.

In 2015 three astronauts were captured on a video crunching red romaine lettuce grown in space. How does a land plant that has been designed by many millions of years of evolution to have its roots firmly anchored in the soil and its shoots rising to the sky grow and reproduce under micro or zero gravity? A team led by Anna-Lisa Paul at the University of Florida investigated this phenomenon for years and found that Arabidopsis plants grown in space altered the expression pattern of their genes, hundreds of them, turning them on or off to adjust themselves to the micro-gravity environment and orient themselves to other environmental stimuli.

Although large-scale space farming seems to be a long way off, it is not inconceivable, considering that plants have adapted to almost every habitat on the earth by reaching down into the inexhaustible genetic reservoir in them and regulating the expression of their genes to adapt to a new environment. It is possible that plants will figure out how to survive in space with their limitless flexibility and plasticity.

While following the news about the *Chang'e-5* mission—China's first lunar sample-return mission in 2020—a piece of news caught my attention: a small packet of rice kernels (about forty grams), along with several other types of seeds, went to the moon with the *Chang'e-5*. After traveling in space for more than twenty days, those rice kernels were brought back and coaxed to germinate on earth. The purpose of this experiment was to expose those rice seeds to high-energy cosmic radiation and microgravity and other possible mutagenic agents in space that could cause genetic mutations in the seeds and possibly result in the creation of mutants with desirable agricultural traits. This process is termed space breeding. Space breeding had been conducted previously in many other space missions, producing thousands of economically useful varieties, including cereal crops, vegetables, and fruits. Although this was the first time that rice seeds traveled to the moon, China had previously sent rice seeds into space through spacecraft and satellites since 1987. So far many mutants have been generated and selected, and more than forty of them have been tested and approved in China as new rice varieties that possess various traits, including high yield, disease resistance, desirable flavor or texture, and even male- or female-sterile lines that could be used to create hybrids.

After seeing this news, I couldn't help wondering if anyone had ever tried to grow rice in space. Indeed, almost forty years ago, the crew in *Skylab*—the first United States space station launched by NASA—conducted hundreds of experiments during its 24-week stay between May 1973 and February 1974. One of those experiments was to study the effects of gravity and light on the growth of rice plants, an experiment proposed by two high school students. Some of the rice seeds sprouted and grew into seedlings; however, because of the small sample size, malfunctioning of the machines, and crudity of the design, this pioneering study generated only rudimentary and inconclusive results.

It is only a matter of time before people will go to Mars or return to the moon. In the following decades we could see more people living and working in space, mining the minerals, building refueling stations, conducting scientific research, or just taking a space adventure. Eventually our future generations may have to colonize Mars or the moon. Once earthlings become a spacefaring species, they need to grow plants in space to sustain themselves and make the Martian air breathable with oxygen released by photosynthesis. As rice has always accompanied us, I am certain that rice will be one of the plants that will accompany us into space, although it may not look the same, just as today's rice bears only a faint resemblance to its wild ancestor and maize looks nothing like teosinte, its former self.

Epilogue

I HAVE BEEN eating rice my entire life, yet I always took rice for granted and never thought of it consciously until I read Malcom Gladwell's *Outliers* years ago. In "Rice Paddies and Math Tests," one of the featured stories in his book, Gladwell argues that rice farming has shaped the culture and values of Asian communities and is the reason why Asians are good at math. After reading the book, some of my Chinese friends thought that Gladwell romanticized and glamorized rice farming, which they believed to be the most excruciatingly painstaking and mind-numbing labor. To search for the truth, I read and collected stories about rice. Over the years the idea of writing something about rice grew on me, but I kept procrastinating.

The COVID pandemic hit. The world shrank suddenly; days merged and became featureless and undistinguishable. During the early days of the pandemic, I felt lost and aimless, so I decided to work on this book. I established a routine, set a goal, and created

EPILOGUE

a sense of purpose. I willed myself to sit at the desk and lose myself in reading and writing. When overwhelmed by the news, events, and chaos around me, I carved my world out of a small desk space, a computer, a stack of books, a cup of tea and narrowed my focus on one thing: rice. Writing was comforting and filled the empty hours. Although I didn't travel during the pandemic, I felt as though I had visited parts of the world—the Po Valley, West Africa, and southern Louisiana. I also went back in time, looking into history and trying to understand where I came from.

My writing progressed at a snail's pace in the midst of waves of the pandemic, the presidential election, Black Lives Matter, Stop Asian Hate, and three missions to Mars. Writing this book was a learning experience more than anything else. As 2020 became 2021 and then 2022, I learned, from the story of rice as well as from the pandemic, that humans are interconnected, whether through the transmission of a deadly virus as is happening now or through sharing culture and food, exemplified by the spread of rice cultivation. A single connection may lead to a string of new connections and then a network of connections. Never before was I so convinced that we are the product of those connections, and it is the reason the human race is still here, suffering, surviving, and prospering.

The experience of the pandemic and writing about rice has expanded my perception and changed the way I think and live. I have become more grateful. I am grateful for the people, past and present, who toiled and sweated to put a bowl of rice on my dinner table and those who have worked day and night to help get the COVID vaccine into my arm. I wouldn't be here without any of them.

A small grain of rice taught me a great lesson during the pandemic.

Acknowledgements

I WANT TO thank my husband, Wenjie Lin, who encouraged me to write this book and told me many stories of his family and his childhood spent in the rice paddy. I want to thank my sons, Justin and Kevin, for their love and support. I also want to thank my parents, my brothers and sister, and all my family and friends who have made my life meaningful and purposeful.

Selected Bibliography

Ahmad, A., et al. 2010. Decoding the epigenetic language of plant development. *Mol Plant* 3(4): 719-28.

Axelsson, E., et al. 2013. The genomic signature of dog domestication reveals adaptation to a starch-rich diet. *Nature* 495(7441): 360–4.

Bak, R. O., et al. 2018. Gene editing on center stage. *Trends Genet* 34(8): 600-11.

Barclay, A. 2007. A hybrid history. *Rice Today* 6(4): 22-5.

Bauer, E. 2021. Indian style rice. *Simply Recipes* July 9.

Becker, J. 1997. *Hungry Ghosts: Mao's Secret Famine*. New York: The Free Press.

Bhavadharini, B. 2020. White rice intake and incident diabetes: a study of 132,373 participants in 21 countries. *Diabetes Care* 43(11): 2643-50.

SELECTED BIBLIOGRAPHY

Biello, D. 2007. When it comes to photosynthesis, plants perform quantum computation. *Sci Am* April 13.

Bogren, R. 2004. New rice variety for crawfish farmers. *Farm Progress* January 15.

Buckley, J. 2021. Sex, Nazis and da Vinci: the hidden history of Italian rice. *CNN Travel* March 5.

Burbank, L. 1907. *The Training of the Human Plant*. New York: Century Co.

Cagan, A., and T. Blass. 2016. Identification of genomic variants putatively targeted by selection during dog domestication. *BMC Evol Biol* 16: 10.

Callaway, E. 2014. Domestication: the birth of rice. *Nature* 514: S58-S59.

Calvino, I. 1983. *Mr. Palomar*. San Diego: Harcourt Brace Jovanovich.

Carey, N. 2012. *The Epigenetics Revolution*. New York: Columbia University Press.

Carney J. 2002. *Black Rice—the African Origins of Rice Cultivation in the Americas*. Cambridge: Harvard University Press.

Carson, R. 1962. *Silent Spring*. Boston: Houghton Mifflin Harcourt.

Chan, S. 2019. Daoist nature or Confucian nurture: moral development in the *Yucong* (*Thicket of Sayings*) in *Dao companions to Chinese Philosophy*. Cham (Switzerland): Springer Nature.

Chan, T. W. K. 2009. Searching for the bodies of the drowned: a folk tradition of early China recovered. *J Am Orient Soc* 129(3): 385-401.

Chang, I. 1995. *Thread of the Silkworm*. New York: Basic Books.

Chang, I. 2003. *The Chinese in America: A Narrative History*. London: Penguin Books.

Char, S. N., et al. 2019. CRISPR/Cas9 for mutagenesis in rice. *Methods Mol Biol* 1864:279-93.

Chen, Y., et al. 2017. Arsenic transport in rice and biological solutions to reduce arsenic risk from rice. *Frontier in Plant Sci* 8:268.

Chen, Z., et al. 2019. Research progress of rice space mutation bio-breeding (in Chinese). *J South China Agric Univ* 40(5): 195-202.

China Daily. 2020. Rice seeds carried to the moon and back sprout. December 29.

Choi, D. 2020. An analysis of Netflix's upcoming 'Water Margin': a difficult adaptation causing both excitement and worry. *Hollywood Insider* December 18.

Chouard, P. 1960. Vernalization and its relations to dormancy. *Annu Rev of Plant Physiol* 11(1): 191–238.

SELECTED BIBLIOGRAPHY

Christoff, P., et al. 2015. Shaoxing wine stories. https://www.academia.edu/17534988/Shaoxing_Wine_Stories_with_a_curriculum_packet_

Cloake, F. 2010. How to make the perfect risotto. *The Guardian* May 6.

Cohen, J. 2004. Mathematics is biology's next microscope, only better; biology is mathematics' next physics, only better. *PLoS Biol* 2(12): e439.

Cohen, J. 2019. The untold story of the 'circle of trust' behind the world's first gene-edited babies. *Science* August 1.

Courville, S. 2016. Crawfish and rice: a winning combination for Louisiana. *Aquaculture North America* March 3.

Cox, T. S., et al. 2002. Breeding perennial grain crops. *CRC Cri Rev Plant Sci* 21(2): 59-91.

Cristopher J. 1957. *The Death of Grass*. London: The Science Fiction Book Club.

Crow, J. F. 1998. 90 years ago: the beginning of hybrid maize. *Genetics* 148(3): 923-8.

Curry, A. 2017. A 9,000-year love affair. *Nat Geo* January 17.

Dai, Y., and C. Liu. 1986. *Fruit as Medicine*. Australia: The Rams Skull Press.

Darwin, C. 1859. *On the Origin of Species*. London: John Murray.

Darwin, C. 1868. *The Variation of Animals and Plants under Domestication*. London: John Murray.

Davidson, E., and M. Levin. 2005. Gene regulatory network. *PNAS* 102(14): 4935.

Demi. 1997. *One Grain of Rice: A Mathematical Folktale*. New York: Scholastic Press.

Deng F., et al. 2018. Engineering rice with lower grain arsenic. *Plant Biotechnol J* 16(10): 1691-9.

Diamond, J. 1997. *Guns, Germs, and Steel*. New York: W. W. Norton.

Dingkuhn, M., et al. 1998. Growth and yield potential of *Oryza sativa* and *O. glaberrima* upland rice cultivars and their interspecific progenies. *Field Crops Res* 57: 57-69.

Dr. Seuss. 1937. *And to Think that I Saw it on Mulberry Street*. New York: Vanguard Press.

Dry, S. A short history of gumbo. *Southern Foodways Alliance* https://www.southernfoodways.org/interview/a-short-history-of-gumbo/

Enserink, M. 2013. Golden Rice not so golden for Tufts. *Science* September 18.

Ermakova, M., et al. 2020. On the road to C_4 rice: advances and perspectives. *Plant J* 101(4): 940-50.

Falchi, M. 2014. Low copy number of the salivary amylase gene predisposes to obesity. *Nat Genet* 46(5): 492-7.

Fan, Y., and Y. Li. 2019. Molecular, cellular and Yin-Yang regulation of grain size and number in rice. *Mol Breed* 39: 163.

Ferl, R., et al. 2015. Spaceflight induces specific alterations in the proteomes of Arabidopsis. A*strobiology* 15(1): 32-56.

Fitzgerald, M. A. 2011. Identification of a major genetic determinant of glycaemic index in rice. *Rice* 4: 66–74.

Flood, A. 2018. Einstein's travel diaries reveal "shocking" xenophobia. *The Guardian* June 12.

Foley, J. 2013. It is time to rethink America's corn system. *Sci Am* March 5.

Fontenot, L. A. 2021 The Hitachi rice cooker in Acadiana. *Lafayette Travel* March 21

Franklin, B. 1770. From Benjamin Franklin to John Bartram. *Founders Online*.

Gajewski, W. 1990. The grim heritage of Lysenkoism: four personal accounts. II. Lysenkoism in Poland. *Q Rev Biol* 65(4): 423–34.

Gerl, E., and M. R. Morris. 2008. The causes and consequences of color vision. *Evol: Educ Outreach* 1: 476–86.

Gladwell, M. 2008. Outliers: *The Story of Success*. New York: Little, Brown and Company.

Graham, S. 2002. Scientists sequence rice genome. *Sci Am* April 5.

Guo, M. 1962. *Qu Yuan*. Beijing: People's Literature Publishing House (in Chinese).

Hagemann, R. 2002. How did East German genetics avoid Lysenkoism? *Trends in Genet* 18(6): 320-4.

Harris, S. R. 2012. We're all in the same boat: a review of the benefits of dragon boat racing for women living with breast cancer. *Evid Based Complement Alternat Med* 167651.

Hedden, P. 2003. The genes of the Green Revolution. *Trends in Genet* 19(1): 5–9.

Hepting, G. H. 1974. Death of the American chestnut. *J For Hist* 18(3): 61–7.

Hibberd, J. M., et al. 2008. Using C_4 photosynthesis to increase the yield of rice—rationale and feasibility. *Curr Opin Plant Biol* 11(2): 228-31.

Hilakivi-Clarke, L., et al. 2010. Is soy consumption good or bad for the breast? *J Nutr* 140(12): 2326S–34S.

Hu, F. Y., et al. 2003. Convergent evolution of perenniality in rice and sorghum. *PNAS* 100(7): 4050-4.

Huang, X., et al. 2012. A map of rice genome variation reveals the origin of cultivated rice. *Nature* 490(7421): 497–501.

Hurles, M. 2004. Gene duplication: the genomic trade in spare parts. *PLoS Biol* 2(7): e206.

SELECTED BIBLIOGRAPHY

Inchley, C. E., et al. 2016. Selective sweep on human amylase genes postdates the split with Neanderthals. *Sci Rep* 6: 37198.

Itkin, M., et al. 2016. The biosynthetic pathway of the non-sugar, high-intensity sweetener mogroside V from *Siraitia grosvenorii*. *PNAS* 113(47): e7619–28.

Jiang, J., et al. 2020. Space breeding in modern agriculture. *Am J Agri Res* 5: 81.

Jiao, Y., et al. 2010. Regulation of OsSPL14 by OsmiR156 defines ideal plant architecture in rice. *Nat Genet* 42(6): 541–4.

Joravsky, D. 1970. The Lysenko Affair. Cambridge (MA): Harvard University Press.

Katz, S. E. 2012. *The Art of Fermentation*. White River Junction (VT): Chelsea Green Publishing.

Kessler, M. 2018. Breast cancer and dragon boat racing: the story behind a movement. *WBUR* November 30. https://www.wbur.org/onlyagame/2018/11/30/sandy-smith-mckenzie-harris-frost

Lazzaris, S. 2019. Rice—the Italian way. *Food Unfolded* July 18.

Le, B. Q. 2018. The story and science of soy sauce. *Science Meets Food* June 6.

Lee, C. 1975. Intense sweetener from Lo Han Kuo (*Momordica grosvenori*). *Experientia* 31(5): 533-4.

Lee-St. John, J. 2009. The legends behind the Dragon Boat Festival. *Smithsonian* May 14.

Lenstra, J. A. 1995. The applications of the polymerase chain reaction in the life sciences. *Cell Mol Biol* 41(5): 603-14.

León-Mimila, P., et al. 2018. Low salivary amylase gene (*AMY1*) copy number is associated with obesity and gut *Prevotella* abundance in Mexican children and adults. *Nutrients* 10(11): 1607.

Li, C. C. 1987. Lysenkoism in China. *J Hered* 78(5): 339-40.

Librado, P., et al. 2021. The origins and spread of domestic horses from the Western Eurasian steppes. *Nature* 598: 634–40.

Linares, O. F. 2002. African rice (*Oryza glaberrima*): History and future potential. *PNAS* 99(25): 16360–5.

Lingan, J. 2015. That old black magic. *Oxford American* 91.

Liu, H., et al. 2018. Genes contributing to domestication of rice seed traits and its global expansion. *Genes* 9(10): 489.

Liu, Y., et al. 2009. Science and politics. *EMBO Rep* 10(9): 938–9.

Lu, X. 2014. *Call to Arms*. New York: Simon & Schuster.

Lucarino-Diekmann, D. 2020. Bella ciao: goodbye beautiful. *La Gazzetta Italiana* September.

Lv, Y., et al. 2019. Current understanding of genetic and molecular basis of cold tolerance in rice. *Mol Breed* 39(12): 159

Ma, Y., et al. 2015. COLD1 confers chilling tolerance in rice. *Cell* 160(6): 1209-21.

Mao Z. 1958. Introducing a Co-operative. https://www.marxists.org/reference/archive/mao/selected-works/volume-8/mswv8_09.htm

McGovern, P. E., et al. 2004. Fermented beverages of pre- and proto-historic China. *PNAS* 101(51): 17593–8.

McKenzie, D. C. 1998. Abreast in a boat—race against breast cancer. *CMAJ* 159(4): 376–8.

Merkys, A. J., et al. 1984. Plant growth, development and embryogenesis during Salyut-7 flight. *Adv Space Res* 4(10): 55-63.

Miller, J. B., et al. 1992. Rice: a high or low glycemic index food? *Am J Clin Nutr* 56(6): 1034–6.

Moss, R. 2018. The real story of gumbo, okra, and filé. *Serious Eats* August 10.

NASA/Marshall Space Flight Center.1973. Plant growth/plant phototropism—Skylab student experiment ED-61/62. https://archive.org/details/MSFC-0102081

Niimura, Y., and M. Nei. 2007. Extensive gains and losses of olfactory receptor genes in mammalian evolution. *PLOS One* 2(8): e708.

Orel, V.1992. Jaroslav Kříženecký (1896–1964), tragic victim of Lysenkoism in Czechoslovakia. *Q Rev Biol* 67(4): 487– 94.

Paine, J. A., et al. 2005. Improving the nutritional value of Golden Rice through increased pro-vitamin A content. *Nat Biotechnol*. 23(4): 482–7.

Pappas, S. 2017. Facts about carbon. https://www.livescience.com/28698-facts-about-carbon.html

Perrett, B. 2008. Qian Xuesen laid foundation for space rise in China. *Aviation Week* January 7.

Perry, G. H., et al. 2007. Diet and the evolution of human amylase gene copy number variation. *Nat Genet* 39(10): 1256-60.

Portal, J. 2007. *The First Emperor: China's Terracotta Army*. Cambridge: Harvard University Press.

Prakash, S. 2019. Why I only use Carnaroli rice to make risotto. *The Kitchn* May 1.

Qian, Q., et al. 2021. Yong Longping and hybrid rice research. *Rice* 14:101.

Qian, X. 1958. What is the maximum grain yield? (in Chinese) *China Youth Daily* July 16.

Qiu, J. 2012. China sacks officials over Golden Rice controversy. *Nature* December 10.

Roll-Hansen, N. 1985. A new perspective on Lysenko? *Ann Sci* 42(3): 261–78.

Sanger, F., et al. 1977. DNA sequencing with chain-terminating inhibitors. *PNAS* 74(12): 5463–7.

Schibler, U., et al. 1982. The mouse α-amylase multigene family sequence organization of members expressed in the pancreas, salivary gland and liver. *JMB* 155: 247–66.

Schubert, M., et al. 2014. Prehistoric genomes reveal the genetic foundation and cost of horse domestication. *PANS* 111(52): e5661-9.

Shao, J. 1997. Research on the origin of Hemudu paddy rice. *Agri Archaeol* 47: 87-92.

Shapiro, J. 2001. Mao's war against nature: legacy and lessons. *J of East Asian Stud* 1(2): 93-119.

Shi, J. 2019. Shaoxing's towpath to history. *SHINE News* March 2.

Shi, N., et al. 2011. *The Water Margin*. Clarendon (VT): Tuttle Publishing.

Shim, J. 2012. Perennial rice: improving rice productivity for a sustainable upland ecosystem. *SABRAO J breed genet* 44(2):191-201.

Shurtleff, W., and A. Aoyagi. 2013. *History of Tofu and Tofu Products (965 CE to 2013)*. Lafayette (CA): Soyinfo Center.

Simon, S. 2019. Soy and cancer risk: our expert's advice. https://www.cancer.org/latest-news/soy-and-cancer-risk-our-experts-advice.html

Singh, R. P. 2011. The emergence of Ug99 races of the stem rust fungus is a threat to world wheat production. *Annu Rev Phytopathol* 49: 465–81.

Skerrett, P. J., and W. C. Willett. 2010. Essentials of healthy eating: a guide. *J Midwifery and Women's Health* 55(6): 492–501.

Slezak, M. 2013. Militant Filipino farmers destroy Golden Rice GM crop. *New Scientist*. August 9.

Smith, J. S. 2009. *Garden of Invention—Luther Burbank and the Business of Breeding Plants*. London: Penguin Press.

Smith, R. J. 1994. *China's Cultural Heritage: The Qing Dynasty*, 1644–1912 (2nd ed.). Boulder: Westview Press.

Song, W., et. al. 2014. A rice ABC transporter, OsABCC1, reduces arsenic accumulation in the grain. *PNAS 111(44)*: 15699-704.

Song, X. 2022. Guilin's geology—a walk on the sea bed. *China Highlights*. https://www.chinahighlights.com/guilin/geology.htm

Sowell, R. Saving of seeds of the world. *Earth Island Journal*. https://www.earthisland.org/journal/index.php/magazine/entry/saving_of_seeds_of_the_world

Spielmeyer, W., et al. 2002. Semidwarf (sd-1), "green revolution" rice, contains a defective gibberellin 20-oxidase gene. *PNAS* 99(13): 9043-8.

Springer, K. 2018. Narezushi: A taste of ancient sushi in Japan. *CNN Travel* January 10.

Stegeman, G. 2020. The history of jambalaya, a true melting pot of flavors. *Chowhound* February 20.

Stevens, G. A., et al. 2015. Trends and mortality effects of vitamin A deficiency in children in 138 low-income and middle-income countries between 1991 and 2013: a pooled analysis of population-based surveys. *Lancet Glob Health* 3(9): e528–36.

Strogatz, S. 2020. *Infinite Powers: How Calculus Reveals the Secrets of the Universe*. Boston: Mariner Books.

Sun, G. 2008. *South China in Ancient Time—Hemudu Site* (in Chinese). Tianjin: Tianjin Chinese Classics Publishing House.

Sun, S., et al. 2018. Decreasing arsenic accumulation in rice by overexpressing *OsNIP1;1* and *OsNIP3;3* through disrupting arsenite radial transport in roots. *New Phytol* 219(2): 641-53.

Sweeney, M., and S. McCouch. 2007. The complex history of the domestication of rice. *Ann of Bot* 100(5): 951-7.

Tai, S. 2019. The battle of renaming rice (in Chinese). https://twgreatdaily.com/jIrPA28BMH2_cNUgaTer.html

Talhelm, T., et al. 2014. Large-scale psychological differences within China explained by rice versus wheat agriculture. *Science* 344(6184): 603-8.

Tang, G., et al. 2009. Golden Rice is an effective source of vitamin A. *Am J Clin Nutr* 89(6): 1776–83.

Tang, G., et al. 2012. β-Carotene in Golden Rice is as good as β-carotene in oil at providing vitamin A to children. *Am J Clin Nutr* 96(3): 658-64.

Tao, D., and P. Sripichitt. 2000. Preliminary report on transfer traits of vegetative propagation from wild rice species to *Oryza sativa* via distant hybridization and embryo rescue. *Witthayasan Kasetsat* 34(1): 1-11.

Toothman, J. 2022. Rice cooker basics. https://home.howstuffworks.com/rice-cooker1.htm

Trefil, J., et al. 2009. The origin of life. *Am Sci* 97(3): 206.

Venkatapoorna, C. M. K. et al. 2019. Association of salivary amylase (*AMY1*) gene copy number with obesity in Alabama elementary school children. *Nutrients* 11(6): 1379.

Viljakainen, H., et al. 2015. Low copy number of the *AMY1* locus is associated with early-onset female obesity in Finland. *PLOS One* 10(7): e0131883.

Wagstaff, K. 2015. One small step for veggies: astronauts eat lettuce grown in space even astronauts have to eat their vegetables. *NBC News* August 10.

Wang, B., and H. Wang. 2017. IPA1: a new "Green Revolution" gene? *Mol Plant* 10(6): 779–781.

Wang, W., et al. 2018. Genomic variation in 3,010 diverse accessions of Asian cultivated rice. *Nature* 557: 43—9.

Wang, Y., and J. Li, 2008. Molecular basis of plant architecture. *Ann Rev Plant Biol* 59: 253-79.

Wild Greens & Sardines. 2010. Paella: in search of the elusive socarrat. October 9.

Wu, S. X. 2021. Yuan Longping (1930–2021)—crop scientist whose high-yield hybrid rice fed billions. *Nature* 595: 26.

Xi, D. 2015. *The Grain that Changed the World* (in Chinese). Beijing: Beijing University Press.

Xiong, M., et al. 2004. Identification of genetic networks. *Genetics* 166(2): 1037—52.

Xu, W. 2017. Yuan Longping announces key breakthrough in stripping rice parents of cadmium. *Yicai Global* September 28.

Yamauchi, T., et al. 2017. An NADPH oxidase RBOH functions in rice Roots during lysigenous aerenchyma formation under oxygen-deficient conditions. *Plant Cell* 29(4): 775-90.

Ye, X., et al. 2000. Engineering the provitamin A (beta-carotene) biosynthetic pathway into (carotenoid-free) rice endosperm. *Science* 287(5451): 303– 5.

Yoffe, E. 1994. Is Kary Mullis God? Esquire 122(1): 68-75.

Young, L. 1988. Regional stereotypes in China. *Chin Stud Hist* 21(4): 32–57.

Yu. J., et al. 2002. A draft sequence of the rice genome (*Oryza sativa* L. ssp. *indica*). *Science* 296(5565): 79-92.

Yuan, L. P. 1966. A preliminary report on male sterility on rice (*Oryza sativa L.*). *Science Bulletin* (English version) 17 (7).

Zhang, Y. 2019. He Jiankui gets three years for illegal human embryo gene-editing. *China Daily* December 30.

Zhao, Y. 2016. An interview with Yuan Longping. *BBC News* (in Chinese) October 4.

Zhao, Y., et al. 2019. miR1432-*OsACOT* (Acyl-CoA thioesterase) module determines grain yield via enhancing grain filling rate in rice. *Plant Biotechnol J* 17(4): 712-23.

Zhou, X. 2002. *Rice—the Important Archaeological Discovery at Hemudu Site* (in Chinese). Hangzhou: Zhejiang Art Publishing House.

Zimmer, C. 2008. Darwin, Linnaeus, and one sleepy guy. *National Geographic* August 19.

The following websites were also used as the sources of information:

Africa Rice Center: https://www.africarice.org/

Britannica: https://www.britannica.com/

Center for Biological Diversity: https://www.biologicaldiversity.org/

SELECTED BIBLIOGRAPHY

C4 Rice Project: https://c4rice.com/

China Discovery: https://www.chinadiscovery.com/

Conrad Rice Mill: https://www.conradrice.com/

Food and Agriculture Organization of the United Nations: https://www.fao.org/home/en

Georgia Historical Society: https://georgiahistory.com/

Golden Rice Project: https://www.goldenrice.org/

Gullah Geechee Cultural Heritage Corridor Commission: https://gullahgeecheecorridor.org/

History: https://www.history.com/

International Rice Festival: https://www.ricefestival.com/

International Rice Gene Bank: https://www.irri.org/international-rice-genebank

International Rice Research Institute (IRRI): https://www.irri.org/

National Geographic Society: https://www.nationalgeographic.org/

New World Encyclopedia: https://www.newworldencyclopedia.org/

Nikon's Small World: https://www.nikonsmallworld.com/

Plants of the World Online: https://powo.science.kew.org/

Rice and Wine: http://www.riceandwine.com/

Rice Knowledge Bank: http://www.knowledgebank.irri.org/

Ricepedia: https://ricepedia.org/

Savannah National Wildlife Refuge: https://www.fws.gov/refuge/savannah/

Science Meets Food: https://sciencemeetsfood.org/
Svalbard Global Seed Vault: https://www.seedvault.no/
World Food Prize Foundation: https://www.worldfoodprize.org/